CHUMASH ETHNOBOTANY

CHUMASH ETHNOBOTANY

Plant Knowledge Among the Chumash People of Southern California

Jan Timbrook

with botanical watercolors by Chris Chapman

SANTA BARBARA MUSEUM OF NATURAL HISTORY
MONOGRAPHS NO. 5
PUBLICATIONS IN ANTHROPOLOGY NO. 1

Santa Barbara Museum of Natural History, Santa Barbara, California
Heyday, Berkeley, California

Library of Congress Cataloging-in-Publication Data

Timbrook, Janice.
Chumash ethnobotany : plant knowledge among the Chumash people of southern California / Jan Timbrook ; botanical watercolors by Chris Chapman.
p. cm. -- (Publications in anthropology ; no. 1) (Monographs / Santa Barbara Museum of Natural History ; no. 5)
Includes bibliographical references and indexes.
ISBN-13: 978-1-59714-048-5 (pbk. : alk. paper)
1. Chumash Indians--Ethnobotany. 2. Ethnobotany--California. 3. Plants--Catalogs and collections--California. 4. California--Social life and customs. I. Title.
E99.C815.T56 2007
581.6'3089975807949--dc22

2006101626

Cover Art: Arroyo Willow (*Salix lasiolepis*), by Chris Chapman
Cover Design: Rebecca LeGates
Interior Design/Typesetting: Philip Krayna Design/PKD, Berkeley, California
Botanical watercolors by Chris Chapman; drawings by Jan Timbrook

Santa Barbara Museum of Natural History
Monographs No. 5
Publications in Anthropology No. 1

The publication of this book was a collaboration of the Santa Barbara Museum of Natural History and Heyday. Orders, inquiries, and correspondence should be addressed to:

Heyday
P. O. Box 9145, Berkeley, CA 94709
(510) 549-3564, Fax (510) 549-1889
www.heydaybooks.com

Printed in Singapore by Imago

10 9 8 7 6 5 4 3

ABOUT THE SANTA BARBARA MUSEUM OF NATURAL HISTORY

The Santa Barbara Museum of Natural History is a private, non-profit institution dedicated to developing and presenting fundamental knowledge of natural history research, so as to teach and inspire a lifelong passion and abiding respect for the natural world.

Founded in 1916, the Museum is housed in a cluster of Spanish-style buildings set on twelve oak-shaded acres in historic Mission Canyon. It features eleven exhibit halls that focus on the Santa Barbara region including birds, insects, mammals, marine life, paleontology, and Native American cultures, as well as special exhibitions, a gallery of antique natural history art, and a planetarium. A satellite facility located on Stearns Wharf offers a window on the Santa Barbara Channel through marine science research, interactive exhibits and aquaria. More than 150,000 people visit the Museum each year.

In addition to presenting public programs and exhibits, the Museum houses internationally significant collections in anthropology, invertebrate zoology, and vertebrate zoology. The Collections and Research division also includes libraries and archives, natural history art, and earth sciences. Ongoing research by curators, students, and visiting scholars contributes to knowledge and understanding of the natural and cultural history of the Santa Barbara region and helps to place significant regional questions into a global context. Collections are open to visitors by appointment. Chumash people serve on the California Indian Advisory Committee and meet regularly with anthropology staff to advise the Museum on collection management, research, public programs, and exhibits related to Chumash culture.

Santa Barbara Museum of Natural History
2559 Puesta del Sol. Santa Barbara, California 93105
805.682.4711
www.sbnature.org

Contents

LIST OF ILLUSTRATIONS

THE CHUMASH AND THEIR NEIGHBORS

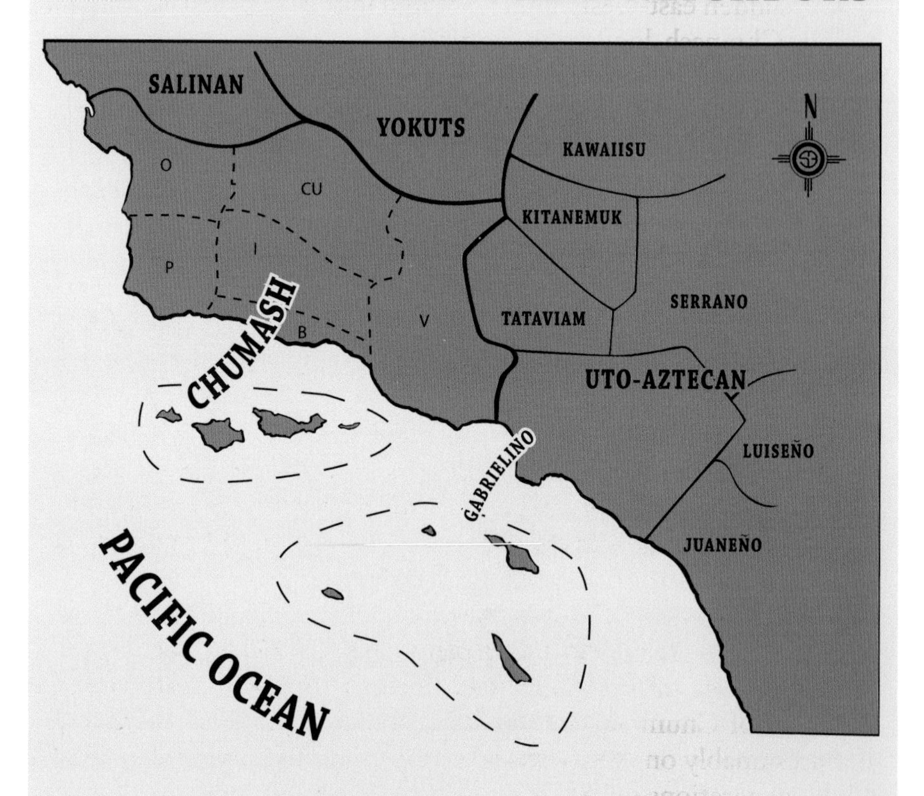

Distribution of Linguistic Groups

CHUMASH		UTO-AZTECAN
O	OBISPEÑO	KAWAIISU
CU	CUYAMA	KITANEMUK
P	PURISIMEÑO	TATAVIAM
I	INESEÑO	SERRANO
B	BARBAREÑO	LUISEÑO
V	VENTUREÑO	JUANEÑO
C	CRUZEÑO	GABRIELINO
SALINAN		**YOKUTS**

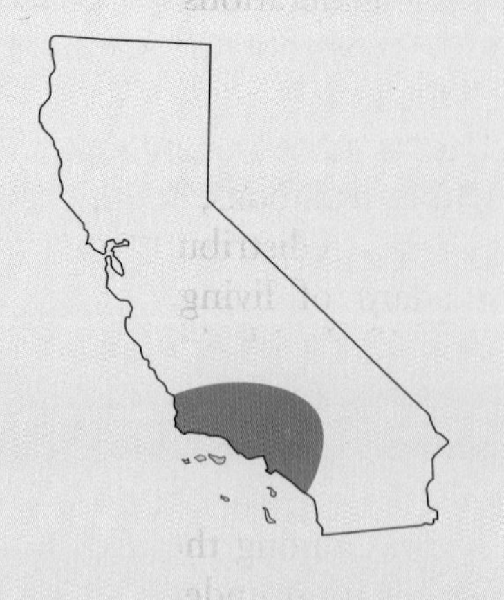

INTRODUCTION

For thousands of years Chumashan peoples and their ancestors have lived where the coastline of southern California makes a sudden east-west bend. Centered in the Santa Barbara Channel region, Chumash lands once extended from Paso Robles to Malibu and from the four Northern Channel Islands nearly to the edge of the Great Central Valley. They included much of what are today Santa Barbara, San Luis Obispo, and Ventura counties.

The Chumash homeland is an area of uncommon biological diversity and richness. The intersection of northern and southern ocean currents in the channel produces upwelling of nutrient-rich waters, creating an immense wealth of marine life surrounding the offshore islands. Along the coastal shelf, low, rolling hills are intersected by stream courses flowing into marshlands and lagoons. Chaparral-covered mountains rise abruptly to 3,500 feet within a few miles of the coast. Farther inland, open river valleys alternate with higher, forested mountains in the backcountry. In a day's walk—or canoe ride—one can reach an enormous variety of habitats, each with its own assortment of indigenous animals and plants. Some fifteen hundred species of plants are native to the Chumash region.

Although the earliest date of human arrival in the channel region has not been established with certainty, there is evidence of occupation on Santa Rosa Island as early as thirteen thousand years ago. The ancestors of Chumash peoples were probably living on the islands—and presumably on the mainland as well—by nine thousand years ago. As the generations passed, their populations grew. They continued to learn more about the local environment and developed new ways of taking advantage of the wealth of natural resources available to them. They adopted or invented new tools for food acquisition (bow and arrow, fishhooks, the plank canoe), and new techniques (leaching, storage, redistribution, and trade) to improve and maintain their standard of living. Thus both their technology and their society diversified with time.

By five hundred years ago, people speaking Chumashan languages numbered about twenty thousand, most living in large, permanent villages along the Santa Barbara coast. In these areas, population density was among the highest in North America. There was no central organization under a single chief. Rather, each village was relatively autonomous politically, though linked to others by social and economic ties. Chumash social structure was affected by wealth, inheritance, and economic specialization—in some ways equivalent to chiefdoms elsewhere. All of this was based on a lifestyle of fishing, hunting, and gathering wild plants. Chumash peoples used fire to promote growth of certain useful plants and to improve hunting, and their gathering practices effectively propagated some plant resources. They did not practice true agriculture or raise domestic animals until the era of Spanish rule beginning in the late eighteenth century. Even then, yields from livestock and planted crops did not notably exceed the former harvest of wild resources.

The impact of Spanish colonization and later American settlement fell heavily on Chumash populations. Nonetheless, through their great resilience and adaptability, these people not only survive to the present day but are thriving throughout the south coast as well as on the Santa Ynez Reservation in the heartland of their former territory.

This book is an attempt to reconstruct as much as possible of traditional Chumash ethnobotany before the coming of Europeans. It also includes information on some plants that have continued to be used well into the twentieth century. This should not be taken to imply that Chumash people no longer use plants, but to recognize that, like any body of knowledge, the role of plants in Chumash life has changed greatly over time.

Purpose of This Volume

Appreciation of Chumashan peoples' lives and cultures requires some understanding of their relationship to the environment in which they lived. Food, medicine, raw materials for making clothing, shelter, all kinds of tools and utensils, religious paraphernalia, and other items essential to existence—all these things were derived in a fairly direct way from the natural world. And, in one way or another, virtually everything Chumash people made and used involved plants. Plants truly were everywhere in Chumash life, not only in daily activities but also in thought, philosophy, and ways of viewing the world. How people *thought about* plants, named and classified them, and wove stories around them—these are just as fundamental a part of culture as how they *used* them.

This is a book about Chumash knowledge and uses of plants, intended for the layperson interested in gaining a fuller understanding of this unique California Indian culture. I hope the work will prove useful to a more scholarly audience and to native people as well.

This is not a book about how you can collect wild plants and use them yourself. Any such attempts are to be strenuously discouraged for two very good reasons. First, there is the issue of the health and safety of the user. There are a disconcerting number of instances of plants being identified incorrectly—in the original sources, by authors, or by would-be plant hunters. Poisonous species may be mistaken for safe ones, and some people may have an individual sensitivity to certain compounds. Use of the wrong plant or even the right plant in the wrong way could result in undesired effects, including death.

Second, and perhaps more important in the long run, we must consider the impact of our actions. Now, in the twenty-first century, there are too many of us and too few natural areas left. The native flora have already been severely persecuted by foreign plants, grazing animals, urbanization, and recreational activities. What wildlife is left truly needs wild plants for survival. We don't. It is possible to appreciate Chumash culture without attempting to emulate it.

A Note on Organization and Language

The plant catalog is arranged alphabetically by botanical name (genus and species). Botanical nomenclature follows that in *The Jepson Manual* (Hickman 1993) and the second edition of *A Flora of the Santa Barbara Region, California* (Smith 1998). Plant names are constantly being revised, however, and botanists familiar with the technical literature will observe that a few names used here may no longer be the official ones. For example, the unfinished *Flora of North America North of Mexico* (Flora of North America Editorial Committee 1993+) has split bulrushes *(Scirpus)* into several genera, changed chaparral yucca to *Hesperoyucca*, and resurrected the Agave family. Because most readers will not recognize these new names, and because the updated ones are not readily accessible, I have decided to stay with the standard reference works most likely to be found on home and library bookshelves.

Common names in English, Spanish, and Chumashan languages are included where known. The three principal Chumashan languages in which plant names were recorded are: (B) Barbareño, from the Santa Barbara area; (I) Ineseño, from Santa Ynez; (V) Ventureño, from the Ventura area. Fewer names are recorded from (Cr) Cruzeño, the island language; (P) Purisimeño, around Lompoc; and (O) Obispeño, from San Luis Obispo. Material from the notes of linguist John P. Harrington is generally discussed first, followed by that from other sources. Where applicable, comparative information from other central and southern California people is given at the end of each entry.

The reader should not be intimidated by Chumash names, which are not impossible to pronounce. Vowels are much like those in Spanish: *a* as in *father; e* as in *then* or *they; i* as in *shin* or *machine; o* as in *long* or *note; u* as in *push* or *rude.* The special symbol *ɨ* represents a sound between short *i* as in *tick* and short *u* as in *tuck;* some have suggested it resembles the sound one makes when stepping barefoot on a slimy slug. The apostrophe represents a glottal stop, like the catch in the throat in *uh-oh*, which can come before the vowel at the beginning of a word, or in the middle or at the end of a word. Consonants are much like those in English, except that *q* represents a sound farther back in the throat than *k.* It is not pronounced *kw* as in quick. The *ch* is like English *chip*, the *sh* like *ship.*

Now I need to apologize to linguists for not using *x* to represent the harsh, breathy sound of *j* in Spanish or *ch* in the German *Bach.* I find that English speakers want to pronounce *x* as *ks*, like the *x* in *fix.* That's to be avoided. Richard Applegate (1975b), whose transcription guidelines I have generally followed here, has suggested an underlined *h̲* to represent this sound. Instead I have used *kh*, which is not completely satisfactory either, but it avoids special diacritical marks that frighten lay readers. Language specialists and others who are interested in more technically accurate transcriptions may consult the list of Chumashan names at the end of this book prepared with the expert advice of Marianne Mithun.

Sources

The principal source of information for this book is the ethnographic and linguistic field notes of John P. Harrington (1884–1961), compiled during his decades of working with Chumash people from about 1911 virtually until his death in 1961. His original notes comprise more than a million pages over all, with about three hundred thousand pages on Chumash groups alone. The notes are housed at the National Anthropological Archives of the Smithsonian Institution. Microfilm copies are now widely available, and the material is being indexed into an online database (Woodward and Macri 2005). My research was conducted in pre-microfilm days, with the original notes and photocopies (see Timbrook 1984, 1990).

Harrington's colorful life and nearly obsessive devotion to recording information from native people, many of them last speakers of barely studied languages, have become legendary and will not be repeated here (see Stirling 1963, Callaghan 1975, Laird 1975, Harrington 1986). His principal Chumash consultants whose knowledge is incorporated into this study were Luisa Ygnacio, Lucrecia García, and Mary Yee (three generations in the same family), and Juan Justo in Santa Barbara; María Solares in Santa Ynez; Rosario Cooper in San Luis Obispo; Fernando Librado, who talked about Santa Cruz Island, Ventura, and Purísima areas; Simplicio Pico and Candelaria Valenzuela, also of Ventura. Harrington also "reheard" material from some earlier work by others, such as Henry Henshaw, who interviewed Juan Estevan Pico in the 1880s.

In addition to taking notes on all possible subjects of language and culture, Harrington also recorded plant names from voucher specimens. Whether it was he or his consultants who collected and pressed the plants is unclear, but some 450 extant specimens are housed at the National Anthropological Archives. Many are labeled with Spanish and Barbareño Chumash common names written on cardboard mailing tags by Lucrecia García and Mary Yee, sometimes with annotations by Harrington. Other tags were written by Harrington's nephew Arthur from interviews with Juan Justo. A few specimens, collected much earlier, are accompanied by slips in Harrington's handwriting with information from Luisa Ygnacio. Although these plant specimens are in precarious condition, some reduced to powdery impressions on the 1923 newspapers in which they had been pressed, most were identifiable to species or at least genus. They made possible the positive botanical identification of many plants mentioned only by common name—in Spanish, English, or up to six Chumashan languages—in the notes. Unfortunately, not every plant found in the notes is represented by an extant specimen, and for some there is just not enough data to make even a guess.

Harrington was a nondirective interviewer; he rarely pursued a linear process of inquiry, other than to clarify points of grammar or pronunciation. So, although I have gone through hundreds of thousands of pages to glean tidbits on ethnobotany, I have found that there are gaps in the information he recorded. On any given page of notes the diversity of topics can be both stunning and frustrating. Say a particular plant was medicinal for fever. What part of the plant? What time of year was it collected? Who gathered it? Was it used fresh or dry?

JOHN P. HARRINGTON
(1884–1961)

LUCRECIA GARCÍA
(1877–1937)

LUISA YGNACIO
(1835?–1922)

JUAN JUSTO
(1858–1941)

MARÍA SOLARES
(1842–1922)

MARY YEE
(1897–1965)

ROSARIO COOPER
(1841–1917)

FERNANDO LIBRADO
(1839–1915)

CANDELARIA VALENZUELA
(1847?–1915)

How was it prepared? What dosage was taken, internally or externally, and how often? He didn't ask.

There are some obvious, inevitable biases. Even though by all accounts Harrington enjoyed close friendships with his Chumash consultants, because he was a male anthropologist he may not have been able to discuss private matters with women. Many of his consultants were elderly, which may explain why he heard about so many medicinal treatments for aches and pains. The entire Chumash geographical range is not equally represented, since consultants knew much more about plants along the south coast and Santa Ynez Mountains than they did about species found farther north or inland.

Still, it's a remarkably complex and detailed body of information. In this book it is augmented with data from other sources, including historical documents from explorers and missionaries and writings by early non-Indian settlers, other anthropological observers, and later researchers. Some archaeological evidence is also incorporated. These other sources supplement and often provide independent support for what Harrington recorded.

It should be pointed out that, just as "Chumash" actually is a catch-all term encompassing peoples spread over a wide geographical region and speaking any of a group of related but in some cases mutually unintelligible languages, so is "Chumash plant knowledge" quite variable between individuals. Some knowledge, such as methods of preparing food from acorns, was quite common and generally consistent from one person to another. Other practices, such as spiritual applications, were far more restricted. To say that "the Chumash" used or thought of a particular plant in a particular way is a bit misleading—in this book, it just means that particular bit of knowledge was reported by at least one somewhat reliable person.

The time frame covers roughly the historic period from initial Spanish contact to the mid-twentieth century. I leave it to others—preferably the Chumash themselves—to compile contemporary plant knowledge and practices.

Acknowledgments

The seeds of this book were planted in the mid-1970s. At that time, Harrington's field notes were just coming to light, revealing extraordinary depth and breadth of information. A flurry of researching, translating, and editing this material resulted in publications on Chumash oral narratives, traditional history and ritual, plank canoe construction and use, cosmology, and more—much of it carried on by the late Travis Hudson, then curator of anthropology at the Santa Barbara Museum of Natural History. As his assistant, translator, and editor, I was inevitably drawn into the process. I'm tremendously grateful to Travis for that opportunity.

In the midst of the Harrington explosion, Robert L. Hoover, professor of anthropology at California Polytechnic University, San Luis Obispo, approached us with a body of Harrington notes that he had discovered on ethnobotany. Because his father was a botanist who had written a flora of San Luis Obispo County, he was very interested in writing up these notes. However, he confessed to having trouble deciphering Harrington's handwriting, being unable to read Spanish, and not being all that familiar with local species of plants. He was looking for a coauthor to help him with the project. Fortuitously, I had already had considerable experience reading Harrington's scrawl, was literate in Spanish, and had a good knowledge of plants thanks to a life-altering undergraduate botany class at the University of California, Santa Barbara. It seemed that this was what I was meant to do. Now, too many years later, I apologize to Bob Hoover for having taken so long to complete his original vision but thank him profusely for allowing me the freedom to do so.

Thanks to research support from the museum, the University of California, and the Smithsonian Institution, as well as the indulgence of Elaine Mills and James Glenn, I was allowed to comb through the original Harrington notes at the National Anthropological Archives in the days before they were organized and published on microfilm. I was also able to examine, photograph, and identify the plant specimens gathered by Harrington's Chumash consultants. Botanists from the Santa Barbara Botanic Garden managed to figure out most of the difficult specimens, for which I thank Ralph Philbrick, Steve Junak, Mary Carroll, and Steve Timbrook. Nancy Vivrette was most helpful with seed identification. J. R. (Bob) Haller first opened my eyes to the plant world. The inimitable Clif Smith, a wealth of information on obscure historical sources and botanical details, was an inspiring and cherished mentor.

So many other colleagues have provided encouragement, support, and necessary prodding. Particular thanks are due to John Johnson, Travis Hudson's successor at the museum; to Thomas Blackburn, one of the original Harrington pioneers; to Avis Keedy and her insatiable curiosity; and to professors David Brokensha, Michael Glassow, Elvin

Hatch, Michael Jochim, and Phillip Walker of the University of California, Santa Barbara. The manuscript was thoughtfully reviewed by Thomas Blackburn, Mary Carroll, Marianne Mithun, and Julie Tumamait-Stenslie. I have done my best to incorporate their valuable suggestions but retain all responsibility for any errors that remain. Thanks also to Chris Chapman for her lovely watercolors that evoke the beauty and diversity of the botanical world, to Kristina Gill for assistance with the index, and to Jeannine Gendar for her unobtrusive but invaluable editing.

I am grateful beyond words to the native people who have generously encouraged this work. Juanita Centeno grew up on the Santa Ynez Reservation and shared plant knowledge gleaned from her Chumash neighbors. Vincent Tumamait and his sister Bertha Blanco shared recollections of their father, Cecilio Tumamait, who was a source for medicinal herbs in the Ventura Indian community. They are gone now, but their trust and friendship are not forgotten. Today, Ernestine Ygnacio-DeSoto, Julie Tumamait-Stenslie, Adelina Alva Padilla, Alan Salazar, Deborah Morillo, Art Lopez, Lei Lynn Olivas Odom, Beverly Folkes, Carmelita Lemos, and others are valued colleagues who are working to make plants an integral part of Chumash life once again.

We are links in the chain of knowledge. From the ancestors to Fernando Librado, María Solares, Luisa Ygnacio, Lucrecia García, Mary Yee, Candelaria Valenzuela, Juan Estevan Pico, Simplicio Pico, Juan Justo, Rosario Cooper; through John Harrington to Travis Hudson, Thomas Blackburn, Richard Applegate, Kathryn Klar, Marianne Mithun, Ken Whistler, and others; through Chumash people today to their children and grandchildren—these many strands of plant knowledge are woven throughout the past, present, and future of Chumash culture.

To the Chumash people—may you always endure.

CHUMASH ETHNOBOTANY

The Plant Catalog

Names for plants in Chumashan languages are designated as:

(B) Barbareño
(Cr) Cruzeño, i.e. Northern Channel Islands
(I) Ineseño
(O) Obispeño
(P) Purisimeño
(V) Ventureño

Spanish language names are denoted as (Sp).

Nonnative plants are labeled "(Introduced)."

Following the botanical name of each plant is the name of or abbreviation for the publishing author who validated that name. This makes it clear what plant is meant, since botanical names sometimes change.

Asteraceae — Sunflower Family

Achillea millefolium L. — **COMMON YARROW**

(B, V) yepunash — (Sp) Yerba Muela,

(I) steleq' 'a'emet — Yerba de la Muela

(I) masteleq 'a pistuk

The Ineseño Chumash names for this plant describe its appearance: "tail of the ground squirrel." The Spanish word *muela* means "molar tooth," and yarrow was used as a medicinal herb for toothache. The plant was boiled and the liquid held in the mouth to reduce pain (Birabent n.d.). Yarrow plants were mashed and applied externally as a poultice for cuts and sores or to stanch the bleeding of a wound (Bingham 1890:36). John P. Harrington's consultants also mentioned the plant being used in this way. Similar uses were noted among the Salinan (Garriga 1978:40) and Ohlone (Bocek 1984:254).

Asteraceae — Sunflower Family

Acourtia microcephala DC. — **SACAPELLOTE**

(B) 'alashkhalalash — (Sp) Sacapellote

Specimens of this plant (formerly *Perezia microcephala*) were collected and identified by Lucrecia García. The Chumash boiled *sacapellote* root and drank this decoction to treat severe or chronic cough. Harrington's consultants mentioned this remedy, and it continued to be used by Chumash descendants into the 1950s (Gardner 1965:299). Earlier sources said that *sacapellote* root decoction was taken for severe colds, lung congestion, asthma, and other pulmonary troubles, as well as applied externally to sores (Blochman 1893a:191; Bard 1894:6; Birabent n.d.). Spanish Californians adopted this remedy as a diuretic to treat kidney and bladder problems (Benefield 1951:22, 25).

The Cahuilla Indians took a decoction of this plant as a sure cure for constipation (Bean and Saubel 1972:99).

Rosaceae — Rose Family

Adenostoma fasciculatum HOOK. & ARN. — **CHAMISE**

(B, I, P) na' — (Sp) Chamiso

The Spanish name *chamiso* most often refers to one particular species, *Adenostoma fasciculatum*, but the word is also sometimes used to mean "brush," that is, unspecified woody vegetation. In this book I apply the term only to *Adenostoma.*

The hard wood of chamise made it useful for a number of tools. In making hollow tubes, the pithy heart of elderberry (*Sambucus*) was reamed or drilled out with chamise sticks. For arrow foreshafts, chamise wood was collected green, then dried by putting it in the groove of a heated soapstone shaft straightener with something like oil to make it smooth. This foreshaft was inserted into the arrow's cane mainshaft (see *Leymus, Phragmites*) and either tipped with a stone point or merely sharpened to a self-point.

Chamise was used for "clam-gathering sticks," rather like a pry bar for abalone (see Hudson and Blackburn 1982:253–254). The hard wood of hinged-stick snares found in the dry caves of the Cuyama area may have been chamise (Hudson 1976:125). Women took a tea made from chamise leaves as a medicine, perhaps for childbirth or menstrual complications.

Other southern California peoples also utilized chamise in various ways. The Cahuilla and Luiseño employed the plant in house and fence construction, for arrows, and as a medicinal wash (Bean and Saubel 1972:30; Sparkman 1908:232).

Rosaceae — Rose Family

Adenostoma sparsifolium TORREY — **RED SHANKS, RIBBONWOOD**

(no Chumash name known) — (Sp) Yerba del Pasmo

This plant has been highly valued as a medicinal, and some authors have suggested that it was exterminated in part of its former range through overexploitation by Spanish Californians (Barber 1931). Another plant, golden fleece (*Ericameria arborescens*), resembles ribbonwood and was sometimes called *yerba del pasmo;* perhaps this extension of the name occurred when the other species became unavailable locally (Timbrook 1987:178). See also *Ericameria. Yerba del pasmo* apparently became more important with Spanish and Mexican colonization than it was to the early Chumash.

Pasmo is a Latin American folk medical condition thought to result from rapid chilling of an overheated body (Kay 1996:53). Dictionaries typically define it as either "spasm" or "paralysis" as in lockjaw. It is not clear what *pasmo* meant to Harrington's Chumash consultants or whether the condition was indigenous to their medical system.

Nowadays the Chumash say that *pasmo* is severe infection or pain caused by air trapped in the body (Antonio Romero, personal communication 1989).

In mission times, *yerba del pasmo* was used for toothache, boiled into a tea drunk to induce sweating, and taken as a powdered snuff like tobacco (Geiger and Meighan 1976:73). Several early authors mentioned its use, both internal and external, to treat tetanus, spasms, and lockjaw (Bard 1894:6; Birabent n.d.; Benefield 1951:21). Harrington's consultants also said that people took *yerba del pasmo* to cause sweating, perhaps as a treatment for paralysis.

Later generations treated toothache, gangrene, and colds with *yerba del pasmo* tea, both drunk and applied to the affected parts (Gardner 1965:297). The tea is still taken as a blood purifier, a treatment for ulcers, sore throats, colds, and disorders of the urinary tract, and used as an external wash for boils and cuts on both humans and livestock. A salve made by mixing the leaves or ground branches with lard or grease is applied to heal sores and cure sore throat; another salve recipe calls for heating *yerba del pasmo* leaves with sulfur and olive oil, then straining the mixture before application (Weyrauch 1982:21)

Polypodiaceae — Fern Family

Adiantum jordanii C. MUELLER — **MAIDENHAIR FERN**

Pellaea andromedifolia (KAULF.) FÉE — **COFFEE FERN**

(no Chumash names known) — (Sp) Culantrillo, Calahuala

Harrington's Chumash consultants knew these two ferns but did not report uses for them. They regarded *culantrillo* and *calahuala* as synonymous names. The use of maidenhair fern in blood disorders and to suppress menstruation was said to be introduced to this area from Mexico (Bard 1894:22).

The Ohlone, unlike the Chumash, distinguished between these two plants. According to information recorded by Harrington, *Adiantum jordanii*, known as *culantrillo*, was taken for stomach troubles, as a blood purifier, and after childbirth; *Pellaea mucronata*, which they called *calahuala*, was used for internal injuries, fever, and facial sores (Bocek 1984:247).

Padre Antonio Garriga of Mission San Antonio de Padua, in Salinan territory, collected herbal remedies used in that region about 1900. Some of his information may have been derived from Chumash practices as well. *Culantrillo* (maidenhair) was boiled and the tea drunk for difficult childbirth, to bring on menstruation, stop hemorrhage, and for the liver; *calahuala* (cliff brake) was used for similar purposes as well as for internal injuries and for the kidneys (Garriga 1978:14, 23, 24, 31–33).

Hippocastanaceae — Buckeye Family

Aesculus californica (SPACH) NUTT. — **CALIFORNIA BUCKEYE**

(no Chumash name known) — (Sp) Berraco

The Chumash seem not to have made use of buckeye. Its natural distribution falls mostly outside their territory, and plants seen in the region today are thought to have been planted or naturalized (Smith 1998:264–265). Chumash consultants knew no native name for this plant, and there were many other species which served the purposes to which buckeye was put by their neighbors.

All groups surrounding the Chumash used the buckeye. The Gabrielino made bows of buckeye for hunting small game. The Yokuts made a special buckeye drill to hollow out arrow mainshafts for insertion of the foreshaft (Latta 1977:292); the Chumash did not need such a drill since their mainshafts were of hollow cane. The Ohlone, Pomo, Mono, and others used crushed buckeye fruits for fish poison (Ambris 1974:80; Mead 1972:6; Munz 1959:994), while the Chumash most often used soaproot (*Chlorogalum*) for that purpose. And the Kitanemuk, like many groups of central and northern California, ate buckeye fruit after carefully leaching out the deadly poison (Harrington 1986; Mead 2003:11–12).

Mission-era Salinans made buckeye fruit into a medicinal tea to bathe hemorrhoids, or mashed it into a paste mixed with beef kidney fat for the same purpose (Garriga 1978:35). The bark and seeds were used medicinally among other tribes (Strike 1994:5–6).

Liliaceae — Lily Family

Agave americana L. (Introduced) — **AGAVE, CENTURY PLANT**

(no Chumash name known) — (Sp) Maguey

This species was an early Spanish introduction from Mexico. *Maguey* and *mescal*, the Mexican names for century plant, were sometimes applied to the local native yucca (see *Yucca whipplei*). During the mission period, Chumash people boiled and pounded agave to produce fibers longer than those of yucca for making string. They used this white cordage for sewing shoes and possibly for basketry handles.

Local medicinal use of agave was also reported. The spines along the edges and at the tip were removed, and then the succulent leaves were split open and applied as a poultice for boils (Birabent n.d.).

Just before flowering, an agave plant can yield one to two gallons a day of sweet juice, which in Mexico is made into fermented and distilled drinks (Jepson 1925:252). This may have been one reason for bringing the plant to California. Harrington's consultants mention a mission priest collecting the juice from century plant stalks, but the Chumash made no alcoholic beverages.

Liliaceae — Lily Family

Allium spp. — **WILD ONION**

(O) tspasɨsɨ — (Sp) Cebolla

"Wild onion" and "*cebolla*" in the literature more often refer to other bulbous roots than to true onions of the genus *Allium* (see *Calochortus*, *Dichelostemma*, and *Zigadenus*).

The Chumash were aware of the strong flavor of wild onions and do not seem to have eaten them very much. Among modern Chumash people, store-bought onions and garlic are sometimes used as an antibiotic to rub on sores (Weyrauch 1982:9–10). This practice may have been introduced from Mexico. Medicinal use of *Allium* appears to have been uncommon among California Indians, though it has been reported as an appetite stimulant, insect repellent, and treatment for snake or insect bites (Mead 1972:8–11; Strike 1994:9).

Betulaceae — Birch Family

Alnus rhombifolia NUTT. — **WHITE ALDER**

(B, I, V) mow — (Sp) Alamillo

Many published references to uses of "alder tree" probably derive from the use of a Spanish dictionary to translate the word *aliso*. In California, however, this term did not refer to alder but to sycamore (*Platanus racemosa*). The Chumash word *mow* was also applied to a kind of sugar called *panocha de carrizo* that was traded by the Yokuts of the San Joaquin Valley (see *Phragmites*).

WHITE ALDER
Alnus rhombifolia

The explorer Fages described Chumash wooden platters made from "roots of oak and the alder trees" (Fages 1937:35), and Harrington's consultants also mentioned alder-wood bowls, trays, and spoons. These spoons were about a foot long, with a straight handle and a round, hollowed bowl like a basket. Bowl-making technology and the guild of woodworkers were described by Hudson (1977). Alder seems to have been less used in woodworking than sycamore and oak.

Alder bark was used as a dye for string. The Chumash soaked the cordage in an alder-bark solution for a day and a night. Alder-bark dye has been reported among some northern California Indian groups (Chesnut 1902:332; O'Neale 1932:27–29), especially as a colorant for woodwardia fern in baskets of the Yurok, Karuk, and Hupa. The Maidu produced the bright orange dye by chewing the bark; the enzymes in saliva may help set the color (Craig Bates, personal communication 1980).

Other tribes used alder-bark tea for medicinal purposes (Chesnut 1902:332–333). Medicinal use of this plant has not been reported for the Chumash.

Amaranthaceae — Amaranth Family

Amaranthus spp. — **AMARANTH, PIGWEED**

(B) sihi' — (Sp) Bledo

(V) sihi

The native *Amaranthus californicus* and several introduced *Amaranthus* species provided seeds which may have been ground and eaten, but the plant was probably of minor importance to the Chumash. Many other Indian groups used either the seeds or the greens, including Mendocino County Indians (Chesnut 1902:346), Wintun (Mead 1972:12), Modoc and Maidu (Strike 1994:11), Cahuilla (Bean and Saubel 1972:37), and Seri (Felger and Moser 1976:19).

Boraginaceae — Borage Family

Amsinckia menziesii (LEHM.) NELSON & J.F. MACBR. var. *intermedia* (FISCHER & C. MEYER) GANDERS — **COMMON FIDDLENECK**

Cryptantha intermedia (A. GRAY) E. GREENE — **COMMON CRYPTANTHA**

(B) tekhe'we — (no Spanish name known)

(I) tekhewe'

Based on plant specimens collected and labeled by Harrington's consultants, these two similar-appearing plants seem to have been classed as one Chumash folk species. The Chumash ate a pinole made of ground and toasted fiddleneck seeds, which they said had a good flavor and a pretty color.

Other California Indians also ate fiddleneck seeds, but cryptantha does not appear to have been used as a food (Heizer and Elsasser 1980:241; Mead 2003:26–27, 138–139; Strike 1994:12).

Primulaceae — Primrose Family

Anagallis arvensis L. (Introduced) — **SCARLET PIMPERNEL**

(B) chikwɨ' — (no Spanish name known)

No early Chumash use has been recorded for this introduced European weed. Although Santa Ynez Chumash lately referred to a specimen of *Anagallis* as "*chikwit*" (Gardner 1965:298), for Harrington's Barbareño consultants *chikwɨ'* meant filaree (*Erodium*).

Adding to the confusion is the similar-sounding English common name "chickweed" usually applied to *Stellaria*, an unrelated plant that was not used.

People remembered scarlet pimpernel being used to treat wounds by boiling the whole plant and tying it on as a poultice over the affected part. Infections were bathed or soaked in the water in which the plant had been boiled (Gardner 1965:298). It has been used by modern Chumash as a poultice or wash for sores and to bathe eczema and ringworm (Weyrauch 1982:8). This practice may have been introduced from Mexico (Ford 1975:173).

Saururaceae — Lizard's-Tail Family

Anemopsis californica (NUTT.) HOOK & ARN. **SWAMP ROOT, YERBA MANSA**

(B, I) 'onchochi — (Sp) Yerba del Manso

(O) ch'elhe' tsqono

(V) 'onchoshi

This important medicinal plant was used both internally and externally. Several early writers noted that *yerba mansa* root tea was valued as a healing wash for cuts, ulcerated sores, and venereal disease (Bard 1894:7; Bingham 1890:37; Birabent n.d.), as a blood purifier (Benefield 1951:21), and as a wash or poultice for rheumatism (Blochman 1894b:40).

Harrington's consultants bathed cuts in *yerba mansa* tea, used it as a hot bath for rheumatism, and drank it for cough. It was considered a good medicine, but too much of it was thought to harm the liver. Plants growing in sandy places were preferred; those in adobe soil were too strong.

Yerba mansa was also used to ritually purify and prepare a person who would be carrying *'ayip*, a poisonous, supernaturally powerful substance made from alum and rattlesnake. In preparation, the person fasted and abstained from sex for eight days, during which time he drank *yerba mansa* tea and washed his face, eyes, and ears with it. For three days before handling the *'ayip*, he cleaned *yerba mansa* roots, chewed them up and swallowed them, and also inhaled the juice squeezed from the smashed roots from a small white clamshell. He was then considered protected from the dangers of working with *'ayip*.

Yerba mansa has continued in use as a wash for infected sores, both in humans and livestock, and the tea is drunk as a blood purifier and

YERBA MANSA
Anemopsis californica

for colds, asthma, and kidney problems (Gardner 1965:297; Weyrauch 1982:22). The root is considered best when picked in the spring and dried before use, though it can be used fresh at any time of year (Weyrauch 1982:22).

This plant is valued medicinally by every Native American people in whose territory it occurs, including all the neighbors of the Chumash (Garriga 1978:31; Bocek 1984:250; Strike 1994:13; Mead 2003:27–28). It is used throughout Hispanic America as well (Ford 1975:341–343). Because of this medicinal importance, humans have intentionally spread *yerba mansa*, and large patches have become established in new locations, such as the California Channel Islands (Smith 1998:347; Timbrook 1987:172–173).

Apocynaceae — Dogbane Family

Apocynum cannabinum L. — **INDIAN-HEMP, DOGBANE**

(B, Cr, I, V) tok — (Sp) Cañasada (?)

Harrington's consultants often spoke of "red milkweed," not a true milkweed (*Asclepias*) but Indian-hemp, which has red stems and milky juice. This was the most important fiber plant of Chumash peoples, and it had many uses in early times.

Indian-hemp was found growing in damp places, including Saticoy, Matilija, Cuyama, and *Sitoptopo* (now Topatopa), a mountain range just north of Ojai. The Chumash collected it in September, though it was in usable condition as early as June. They either cut the stalks or pulled them up; they preferred to pull them, as it was less wasteful of usable material. They cut off the top part and the leaves and brought home just the stalks, or they might do further processing right at the collecting site. The island Chumash had no Indian-hemp locally available, but they acquired it from the mainland people in exchange for sea otter skins.

To prepare the *tok* fiber, the Chumash dried the stalks and then pounded them with a stick or rolled them on stones (Henshaw 1955:152). They then removed the woody parts and rubbed the remaining fiber between the fingers to get the outside scale off, shaking out everything that was not useful. They shredded the fiber and then half twisted it by rolling it on the thigh with the palm of the hand. They tied lengths of half-twisted fibers together in bunches.

Harrington reported that both men and women made string (1942:25). They placed two half-twisted fibers parallel on the thigh, the end of one laid even with the tip of the other. They twisted them further by rolling both at once toward the knee. Then they rolled them together back in the opposite direction, toward the body, to unite them into one two-ply cord. This process worked best if one moistened the hand in a dish of water or by spitting on it. On reaching the ends of the strands, more fibers were spliced on, so that the cord could be made any length. To keep the cord from unraveling, a knot was tied in the end.

Two-ply string was the most common, but two or three two-ply strands could also be twisted together for thicker, stronger cordage. The Chumash also made three- and four-strand braid. To test the strength of the cord, they held the ends behind the hands and gave the strand three sharp jerks. They wound the finished string around the arms, straightened it, and stored it in skeinlike coils until needed. They did not roll the string into balls.

The Chumash employed Indian-hemp fiber string as a sewing thread for buckskin bags, as bowstrings, as lashings for plank canoes, and for all kinds of fishing equipment. They considered it the best fiber for this purpose since the action of the water hardens it. For canoe-making, this cordage was usually tarred.

Fishing lines were of either two-ply or three-ply Indian-hemp cordage, 120 to 180 feet long. Naturally reddish-brown in color, this cordage was already less visible underwater than other fibers, but the Chumash sometimes colored it with red ochre. This same cordage was also used for bait lines, towed behind the canoe when paddling fast, to draw large fish in close to the canoe so they could be harpooned. At the end of the bait line was a decoy made of abalone shell and a float made of fish air bladder. The harpoon line was three two-ply strings twisted together. It could be as long as 360 feet and was kept coiled up in a shallow basket in the canoe.

The Chumash fished with various kinds of nets that they made with *Apocynum* cordage. Drag nets for towing behind canoes were unweighted and six to eight feet long. They were about eighteen inches wide at the mouth, which was kept open with a four-inch-wide band of weaving, and wider at the opposite end. River nets for catching salmon were also made with Indian-hemp.

This fiber was also used to make carrying nets and bags. The Chumash collected *islay (Prunus ilicifolia)*, acorns (*Quercus* spp.) and manzanita berries (*Arctostaphylos* spp.) in net bags of woven *tok* about eighteen inches wide and three feet long, with a one-foot diameter hoop of sumac (*Rhus trilobata*) at the opening. They made bags with finer meshes to carry shell beads and pine pitch. These bags were very strong and lasted for many years.

For carrying loads of firewood and bundles of plant material, three-ply twisted *Apocynum* fiber string one-quarter inch in diameter was made into a tumpline. One carried the load on one's back suspended from the tumpline, which passed over the forehead and was padded either with a cushion or with a wider, woven forehead band. Slings, most likely an implement introduced by the Spanish, were made of Indian-hemp fiber. People standing on platforms at Ventura Mission used these slings to kill or scare off birds from the fields.

Other articles made from this plant included bands woven or braided of two-ply cordage for lacing babies into cradles, often with shell or glass bead decorations; bracelets with beads sewn on; belts and headbands; and necklaces of strung beads or ornaments.

The Chumash made all kinds of ceremonial paraphernalia and dance regalia with *tok* fiber cordage. The feathered headband was a band of woven *tok* string two and a half inches wide, covered with feather down glued on with tar. The dancer wrapped this around his or her head before donning the feather crown, a framework of wooden hoops fastened together with *tok* string and covered with feathers inserted into twisted cordage. The Fox Dance headdress had a long tail braided of tule and Indian-hemp, and a rock tied to the end of the tail with hemp. Two kinds of dance skirts incorporated *Apocynum* cordage. One was a network to which feathers were attached; the other had a fringe of suspended cords into which feather down had been twisted. The Chumash made ceremonial feather banners similar to those of other California peoples by laying the feather quills parallel to one another and weaving (or stringing?) them with a thread of Indian-hemp. Pieces of *tok* stem sometimes served as *peón* gamesticks.

Indian-hemp appears in Chumash myths, like the following story told to Harrington by Juan de Jesús Justo (see also Blackburn 1975:210):

> Coyote went to *sitokto'k* and bought a great load of *tok*. By his magic power he was able to put it all into his carrying net and bring it all home at once...Next morning Coyote began to work with the *tok*. He worked very fast, twisting it on the machine that he had: his thigh. Coyote twisted the *tok* into string and made fishlines for all the important people of the village.

This species was important throughout western North America (Hoover 1974:7–9). Nearly all California tribes used it, and those in whose territory it did not grow obtained it through trade (Mathewson 1985:17). Many used fire management to produce the almost unimaginable quantities of these stalks required for cordage and nets (Anderson 2005).

Ericaceae — Heath Family

Arctostaphylos spp. — **MANZANITA**

(B) sq'oyon — (Sp) Manzanita

(I) sq'o'yon

(V) tsqoqo'n

Like the Spanish name *manzanita*, or "little apple," the Chumashan names refer specifically to the fruit rather than the whole plant. Manzanita specimens collected by Harrington's Chumash consultants are mostly bigberry manzanita (*Arctostaphylos glauca*), but other species were probably used as well.

Chumash people gathered manzanita fruits in summer and dried them, then ground them on a metate and ate them in winter as a coarse meal. Sometimes they ground up the fruits while still fresh. It is not clear whether they ate the dry outer pulp and skin, or if it was the seed itself that was ground into meal. They did say that the mashed, dried berries were winnowed in abalone shells or baskets; this may indicate that the skins were discarded. The Chumash ate the ground-up berries raw as pinole, mixed with a little water or, in historic times, with milk. Like some other California peoples, they made a beverage of manzanita, although they put branch tips as well as the fruits into the water for a pleasant drink.

A few other, nonfood uses for manzanita are also known. Dried fish were smoked over a fire of manzanita wood. Among modern Chumash, the water in which manzanita berries have been boiled is occasionally used as a treatment for poison-oak rash (Weyrauch 1982:12).

Arctostaphylos spp. were used by Indian peoples throughout California (Strike 1994:17–19; Mead 2003:37–45). The berries were eaten fresh or they could be dried and pounded into flour to be eaten raw, cooked into mush or biscuits, or made into a beverage. A few groups used the berries and leaves medicinally; some smoked the leaves in pipe mixtures; and various peoples used the wood for pipes and construction material. The Cahuilla preferred Eastwood manzanita (*A. glandulosa*) over *A. glauca* for eating fresh (Bean and Saubel 1972:40–41).

MANZANITA
Arctostaphylos sp.

Asteraceae — Sunflower Family

Artemisia californica LESS. — **COASTAL SAGEBRUSH**

(B) we'wey — (Sp) Romerillo

(Cr, I) wewey

(O) tilho

(V) wewe'y

Coastal sagebrush, also called California sagebrush, is distinct from true sage (*Salvia* spp.). The Chumash used sagebrush in a wide variety of ways, including tools, construction, medicine, and ritual. They kindled fire with firesticks consisting of a horizontal hearth piece of sagebrush and a vertical drill of *guatamote (Baccharis salicifolia)*. They sometimes made foreshafts of composite arrows from sagebrush, which they considered to be very strong. The foreshaft was about six to nine inches long, straightened with fire, and had a sharp, fire-hardened point. It was inserted into the *carrizo* mainshaft, glued with tar, and secured with sinew binding. Barbed foreshafts could be made from a piece of sagebrush that had a branch starting off at an angle.

The Chumash constructed windbreaks and brush barricades around dance grounds with *romerillo* branches. They sometimes burned the plant as fuel for cooking. It was said that some old men used to smoke the leaves for recreational purposes.

This species had several medicinal uses. Headache was treated by tying the leaves on the forehead (Birabent n.d.) or by rubbing the leaves between the hands in water which was then used to wet the hair and head. A person might also carry the leaves inside the hat; it made the head feel good and cool. People suffering from paralysis could inhale the steam from boiling *romerillo* leaves, which was thought to be an effective treatment. One of several reported poison-oak remedies was made from water in which *romerillo* leaves were boiled (Birabent n.d.). More recently, Chumash people have used the plant to treat coughs (Gardner 1965:299).

Romerillo also played a role in ritual. Some of the following traditional uses of *Artemisia californica* reported by Harrington's consultants are still practiced by Chumash and other native peoples today, but using white sage (*Salvia apiana*) instead.

To bless a new pipe, three old men came and sprinkled water in the four directions with a sprig of *romerillo*. (In Harrington's notes, "old men" can just mean old-timers, but the term often refers to respected elders knowledgeable in religious or ritual practices.) When someone had died and the corpse was still in the house, great quantities of the branches were placed around the house as a disinfectant. People burned the plant as an incense to fumigate the house after the funeral, and also used it as fuel to burn the belongings of the deceased five days after burial. The relatives purified themselves by washing their faces and hair in water in which sagebrush had been soaked.

There was a sacred place on the coast west of Ventura where people would take the possessions of the dead in carrying nets and deposit them. In going to this place, the bearers would take along small bundles of sagebrush and place one on the ground about every thirty or forty steps on their ascent, until they arrived at the spot where the goods were to be left. Fernando Librado thought that this practice may have been a later substitute for the old-time custom of burning everything on the fifth day after burial.

Another use of *romerillo* in connection with death was the following: when a widow or widower feared seeing a vision, an old man filled his stone pipe with pounded *pespibata* (see *Nicotiana*) and *romerillo* and blew smoke over the person's head and down the whole body. Then he tied *romerillo* sprigs into a cross about one foot across and two inches thick. This was covered with a skin and placed under the head of the patient, who then slept untroubled by visions. María Solares said this practice dates from before the Chumash became Christians.

During the winter solstice ceremony, bundles of *romerillo* were erected as offerings at shrines. These bundles were three feet long and ten inches in diameter. They were merely branches of *romerillo* arranged with the tips all pointing in the same direction and tied in the middle with a strip of willow bark (see Hudson et al. 1977:61–62).

Asteraceae — Sunflower Family

Artemisia douglasiana BESSER — **MUGWORT**

(B, I) molush — (Sp) Estafiate, Yerba Ceniza

(Cr) qloqol

(O) tpinusmu'

(V) moləsh

Chumash doctors practiced a technique called *'apin*, or cauterizing, which was regarded as one of the most important remedies available. To perform this treatment, they rubbed dry mugwort leaves between the palms to remove the leaves' chaffy skin and leave the fine, woolly hairs. They then pressed this "wool" tightly with a twisting motion, using the fingers of one hand against the palm of the other, to form it into a cone shape. It took two or three handfuls of leaves to prepare each cone. Sometimes they chewed a little tobacco and added the spittle to aid in shaping the cone. Depending on the seriousness of the illness and the strength of the patient, the doctor applied one or several cones to the affected area, an inch or two apart, and lighted them all at the top simultaneously. The cones burned slowly and evenly with an aromatic smoke. At first they did not hurt the patient at all, but they began to get hot after burning halfway down. The treatment left shiny scars about the size of a dime on the skin, and old people often had many of them on their legs and other parts of the body.

This was the preferred remedy for wounds, skin lesions, and rheumatism because, as María Solares said, "it kills your infirmity, your ache." In a variation on the technique, the doctor might heat a stone tube or pipe by burning dried mugwort leaves inside it, and then apply it to the diseased spot to induce a blister (Bard 1894:9). Both these cauterizing practices utilized a counterirritant principle similar to the "cupping" of eighteenth- and nineteenth-century European medicine. Another species of *Artemisia* has been similarly used as *moxa*—burned on or near the skin—in Asian medicine since ancient times.

Mugwort had other medicinal uses as well. For sore muscles, the leaves were applied as a plaster. In the treatment of *pasmo* (a condition caused by rapid chilling of an overheated body), sweating was induced by laying a thick layer of mugwort leaves over hot coals to form a mat for the sick person to lie on. In recent decades, Chumash people have applied these leaves as a poultice for headache or toothache, boiled them into a tea drunk for headache and asthma, and used them as a soothing bath for measles (Gardner 1965:298; Weyrauch 1982:9).

Today, the most common use of mugwort is as a remedy for poison-oak rash. The Chumash appear to have been less allergic to poison-oak (*Toxicodendron diversilobum*) in earlier times, but some did suffer from its effects. Harrington's consultants mentioned two ways in which mugwort could be used to treat poison-oak. One could boil the leaves and bathe the skin with the resulting tea for several days. Or one could rub the crushed leaves directly on the skin, sometimes with olive oil; this was reported to cure after three or four applications. A third treatment reported more recently is to place the leaves in a cloth bag, crush it, and rub the bag on the skin (Weyrauch 1982:9).

A Barbareño Chumash myth suggests that baby cradles were formerly lined with mugwort leaves: Luisa Ygnacio told Harrington that when Coyote's twenty-four children were born, he put mugwort on their cradles and over them. "It was clothing of babies in the old days," she said.

Elsewhere in southern California, mugwort was used for cauterizing sores and burned as an incense to treat the sick; it also played a part in both boys' and girls' puberty rituals (Boscana 1933:46, 48, and annotations by Harrington). The Salinan used the leaves to treat burns and infections and to remove spots from the face (Garriga 1933:20, 27, 39). Many other California tribes had uses for this plant (Mead 2003:48–50).

Poaceae — Grass Family

Arundo donax L. (Introduced) — **GIANT REED**

(V) shukepesh 'ishaq — (Sp) Carrizo de Castilla

"*Carrizo*" is a term also applied to the large native grasses, *Phragmites* and *Leymus*; "*de Castilla*" refers to the plant's having been introduced by the Spaniards. *Arundo* originated in the warm regions of the Old World but is now widely naturalized from early plantings of the Spanish and Mexican periods (Smith 1998:106). It looks rather like bamboo, and in fact Harrington's notes sometimes refer to it as "European bamboo."

Arundo was a favorite plant in mission gardens. It was used for living hedges and windbreaks, as it grows in thick clumps and will branch if the stalks are cut down. It was widely employed in the construction of adobe houses, lattices, mats, and screens in Hispanic America (Hitchcock 1950:186, 190; Padilla 1961:25). Old photographs of Spanish missions show that early adobe buildings had ceilings of *Arundo* under the fired roof tiles (Smith 1998:106; Webb 1952:87). The sacristy at Mission San Luis Obispo and the ranch house at Los Ojitos near Mission San Antonio de Padua had ceilings of this type (Webb 1952:125–126). Harrington's consultants said that the ranchería buildings at Cieneguitas, near Goleta, were also roofed with *carrizo*.

Arundo was used for many of the same purposes as native species with similar properties of hollow, lightweight stems. For example, the Chumash made arrow mainshafts from the strong side shoots of *Arundo*, considering them to be lighter than the native *Phragmites* stalks. About twenty arrows of this type could be carried in a quiver. Knives were made by splitting the thickest *Arundo* stems and scraping them to create a sharp edge and a pointed end. These knives were an effective tool for cutting meat and gutting small animals. Another implement was the rock thrower, a hollow stick about two feet long used to hurl stones in sham fights between men and boys. The Chumash used elderberry stems (*Sambucus*) for this purpose in earlier times.

Apparently *Arundo* provided the missionized Chumash with a common substitute for elderberry wood in musical instruments. Flutes, made originally of elderberry but later of *Arundo*, were about two feet in length and had four holes or stops. The split-stick clapper, also of either elderberry or *Arundo*, was a section of stem which was split lengthwise, leaving a portion unsplit for the handle. *Arundo* main stalks were thick, and sometimes one would have to cut away a portion of the sides at the base of the split to make it shake better. This instrument was held in one hand and struck on the palm of the other hand to accompany singing with a clapping sound. The musical bow, not originally used by the Chumash, was introduced by Yaqui Indians and others from Sonora, Mexico. It consisted of a single string stretched lengthwise on a piece of *carrizo* stem an inch thick. To play a tune, the player plucked on the bowstring while holding one end of the bow to his mouth, which acted as a resonating cavity.

Asclepiadaceae — Milkweed Family

Asclepias eriocarpa BENTH. — **BROAD-LEAVED MILKWEED**

Asclepias fascicularis DECNE. — **NARROW-LEAVED MILKWEED**

(B, I) 'okhponush — (Sp) Jumete [?]

(V) 'usha'ak

Milkweed was an important fiber plant of the Chumash, second only to Indian-hemp (*Apocynum*) in value. The fact that Harrington's consultants referred to both plants as "milkweed" has caused some confusion. Both *Asclepias* and *Apocynum* ooze milky white juice when cut, but the stem fibers differ in color. Hence *Asclepias* was known as "white milkweed" and *Apocynum* as "red milkweed," even though they belong to different plant families. Of the true milkweeds, Chumash consultants distinguished between *Asclepias eriocarpa*, which has broad, ashy-colored leaves, and *A. fascicularis*, which has narrow, green leaves.

DRESSED TO DANCE

Two members of the Chumash 'antap *cult prepare a young initiate for a ceremonial dance. He wears a down-covered headband and a crown of split feathers topped by magpie tails. His fringed skirt is made from milkweed string twisted with eagle down. Dance wands and body paint complete the regalia. The two men are dressed for the festive occasion, but they will not take part in the dances until the stars of the Pleiades appear late in the night.*

Both men and women made cordage. They cut *Asclepias* stems while still green, allowed them to dry, and then rubbed them to extract the stem fibers. These they twisted into string in much the same way they processed *Apocynum.* The Chumash regarded *Asclepias* as the weaker of the two fibers. They kept its limitations in mind when making articles of cordage and used it most where *Apocynum* was scarce.

The Chumash used carrying nets and tumplines of milkweed fiber only for carrying small things, not heavy weights. They sometimes made nets for gathering acorns and *islay (Prunus ilicifolia)* fruit from *Asclepias* cordage; these were of fine mesh and had a wooden hoop at the mouth so they would stay open. They also made cradle bands and men's belts of this fiber. For the dance apron, they twisted white feather down with white milkweed fiber into cordage, strands of which hung from a belt down to knee length with feathers suspended from the ends.

Milkweed juice was reportedly made into chewing gum. One cut the young plants to drain out the sap and allowed it to congeal. This gum was bitter, but if one spit frequently while chewing, the bad taste soon disappeared.

A plant called *jumete*, apparently a species of *Asclepias*, was used as a purgative. Harrington's consultants also said that if you wanted to hurt anybody—for example, if you had a grudge against someone and wanted revenge—you could put the juice of *jumete* into that person's food or drink in order to make him sick; this was mentioned by other sources as well (Birabent n.d.). The Ohlone used stem fibers of a plant they called *jumete*, which has been identified as *Asclepias fascicularis*, for making cordage (Bocek 1984:252).

A number of other California peoples used *Asclepias* for cordage and fiber articles, and some also chewed the congealed sap (Bean and Saubel 1972:43–44; Jepson 1939:109–111; Heizer and Elsasser 1980: 242; Hoover 1974:7–8). Although many milkweeds are toxic to livestock and all should be regarded with caution, no human poisonings have been reported for these species (Fuller and McClintock 1986:83).

Chenopodiaceae — Goosefoot Family

Atriplex spp. — **SALTBUSH**

(B) 'i'laq' — (Sp) Barbazo, Matorral Grande

(V) mo'

Euphorbiaceae — Spurge Family

Croton californicus MUELL. ARG. — **CALIFORNIA CROTON**

(B) 'i'laq' — (Sp) Barbazo, Matorral Chico

(V) lilak

Some *Atriplex* species and crotons are similar in appearance, despite being members of different botanical families. Specimens collected by Harrington's Chumash consultants indicate these two plants were regarded as basically the same but distinguished by size. Although the notes are unclear, both plants seem to have had similar uses. (See also *Croton* under separate listing.)

Atriplex lentiformis, common along coastal bluffs between Santa Barbara and Ventura, was the source of the village name *Simo'mo*, which means "the saltbush patch" (Applegate 1975b:41). Harrington's consultants said the Indians had a soap factory at *Simo'mo*, inland from Mugu, where they burned croton and used the ashes as lye for making soap. Croton or *Atriplex* was also burned for this purpose at Mission Santa Barbara.

It is doubtful that this form of soapmaking would have been practiced before Spanish colonial times. Soap was made at the missions using lye made from wood ashes; sometimes potassium carbonate obtained from burning seaweed or other plants was added to the lye to increase its efficiency (Webb 1952:203).

The plant known as *barbazo* was made into a tea for colds, to be drunk hot at bedtime (Birabent n.d.). Croton was used in this way by the Cahuilla, but they also used *Atriplex* flowers, stems, and leaves to treat colds (Bean and Saubel 1972:56).

The Cahuilla made soap of *Atriplex* spp. by pounding the roots with water or by rubbing the leaves into articles to be cleaned. They also ground *Atriplex* seeds and ate this as mush or bread (Bean and Saubel 1972:45). The Chumash do not seem to have used the plant as food.

Poaceae — Grass Family

Avena spp. (Introduced) — **WILD OAT**

(B) 'aluche'esh — (Sp) Avena

(I) shushtewesh

These grasses were introduced early in the mission period and quickly adopted by the Chumash as a food source, which they prepared in much the same way as native plant seeds. They picked wild oats by bending the tops over the mouth of a basket and removing the seeds with the hand or a seedbeater, and used a mano with a flat rock to remove the outer chaff. They pounded up the seeds in a mortar and mixed them with ground chia (*Salvia columbariae*) and water for a hot-weather drink. No doubt, several kinds of native grasses were also collected and used in this way, but insufficient information was provided by Harrington's consultants to enable determination of species. Chumash people also cut the whole plants of wild oats for use as livestock fodder. The Ineseño name means "sharp, a spine that digs into your skin."

Asteraceae — Sunflower Family

Baccharis pilularis DC. — **COYOTE BRUSH, CHAPARRAL BROOM**

(no Chumash names known) — (Sp) Rama China

Residents of the Santa Ynez Reservation in the late 1950s considered this plant to be the best remedy for poison-oak rash. They boiled the leaves and applied them to the affected parts (Gardner 1965:297). Juanita Centeno said the branchlets were used to brush away the tiny spines when collecting prickly-pear cactus fruit (personal communication 1978). The Spanish name, *rama china*, or "curly branch," probably refers to the crinkly leaves.

Asteraceae — Sunflower Family

Baccharis plummerae A. GRAY — **PLUMMER'S BACCHARIS**

Conyza canadensis (L.) CRONQ. — **HORSEWEED**

(B) wili'lik' — (Sp) Yerba del Aire, Yerba del Aigre

(I) wililik'

(V) wililik

Plant specimens collected by Harrington's consultants indicate that the Chumash included both these species in the folk category *wililik'*. This herb had medicinal uses: it was ground and rubbed on any aching part of the body, and a tea made from it was drunk for kidney trouble.

The Ohlone applied the name *yerba del aigre* to a different plant, vinegar weed (*Trichostema lanceolatum*), which they used in the treatment of pain, sores, stomachache, and colds (Bocek 1984:253).

Asteraceae — Sunflower Family

Baccharis salicifolia (RUIZ LOPEZ & PAVON) PERS. — **MULE FAT, WATER WALLY**

(B, I) shu' — (Sp) Guatamote

(O) twalilí'

(V) wita'y

This plant was a favorite for making firesticks. The vertical stick, or drill, was as thick as a lead pencil. Its tip rested in a small, cup-shaped depression in the horizontal hearth stick, and the drill was rotated between the palms. The resulting friction produced hot sawdust, which would be caught on dry tinder or punk and then blown into a flame. According to Harrington's consultants, the vertical drill stick was usually made of *guatamote* and the horizontal hearth of *romerillo* (*Artemisia californica*), though some said the hearth was of *guatamote* and the drill of elderberry wood (*Sambucus mexicana*).

The Ineseño Chumash made fish traps called *wisay* to take advantage of large runs of steelhead (and possibly salmon), which were fairly common in the Santa Ynez River. First they piled a barrier of stones across the stream. They cut *guatamote* stems and stripped them of leaves, twisting two or three of these stems together to form a hoop about two and a half feet in diameter. They attached more stems around the edge of the hoop and tapered them together to a point at the opposite end. Several rows of twined weaving, much like that used in making mats, held the stems together. The result was a conical, openwork fish trap. They placed these traps in the river at gaps they had left in the stone dam, and tied them to the larger rocks to hold them in place with the traps' mouths turned upstream. People entered the pools above this and gradually drove the fish downstream into the traps. *Guatamote* stems were also occasionally used to make fishing poles and arrows.

This species is often mentioned in the literature under the names *Baccharis viminea* or *B. glutinosa*, both now considered synonymous with *B. salicifolia* (Hickman 1993:210). Other California tribes used the plant in ways similar to the Chumash. It was a favorite construction material of the Cahuilla (Bean and Saubel 1972:46). Closely related species were used by the Pomo for arrow shafts and by the Luiseño for fire drills (Mead 1972:33). Medicinal uses for the plant are reported among some of these same groups, but not by the Chumash.

Brassicaceae — Mustard Family

Brassica spp. (Introduced) — **MUSTARD**

(no Chumash names known) — (Sp) Mostaza

All the local species of mustard (*Brassica* spp.) are introduced weeds from Europe which became naturalized during the mission period. The Chumash used the young plants as greens. Wild mustard greens were also eaten by the Yokuts, Luiseño, Pomo (Mead 2003:75–76), and Cahuilla, who ate the seeds as well (Bean and Saubel 1972:47).

California folklore has it that seed of black mustard (*Brassica nigra*) was scattered by the earliest missionaries on their way northward to San Francisco, so that the expedition could find its way back again along the same route the next year by following the yellow flowering plants. Doubt is cast on this popular legend by the fact that early botanical explorers who came afterward do not mention this plant, and by the fact that the plants are in flower for a relatively brief season of the year (Jepson 1936:49).

Portulacaceae — Purslane Family

Calandrinia ciliata (RUIZ LÓPEZ & PAVÓN) DC. — **RED MAIDS**

C. breweri WATSON

(B, I, V) khutash — (Sp) Pil

These spring wildflowers produce quantities of small, shiny black seeds that the Chumash harvested and stored in large baskets in their houses. As needed, people would take out small amounts, and toast and grind them for food. When pounded, the seeds made an oily dough which was easily gathered up into balls with the fingers.

The Chumash may have gathered red maids seeds in the same way reported among the Miwok, who pulled up the whole plants in late spring and spread them out on swept ground or on rock outcrops to dry. They beat the plants with sticks, then picked them up and shook them, leaving the dislodged seeds to be swept up and winnowed to remove dirt and chaff (Barrett and Gifford 1933:152–153). The Chumash practice of burning grasslands on a regular basis probably assured an abundant supply of red maids, as this plant is a known fire follower (Timbrook, Johnson, and Earle 1982).

Calandrinia was a highly esteemed food and is mentioned in Chumash myths. One story describes Coyote arriving in a village of strangers. He pointed to the shiny black pupil of his eye—which resembled a red maids seed—and thus made them understand that he wanted *khutash* to eat. When they brought it to him he began pounding it up, nibbling a little now and then, so that by the time he had it all pounded he had eaten it all (Blackburn 1975:207).

Along with acorns, chia seeds (*Salvia columbariae*), and *islay* kernels (*Prunus ilicifolia*), *pil* was one of the most expensive foods in Chumash culture. All these were much sought after in trade and were measured in the standard unit of volume, the woman's basketry hat. Island Chumash people reportedly came to the mainland to buy hatfuls of these seeds, as much as they could carry.

Other groups besides the Chumash considered these seeds to be a delicious food. Harrington's Kitanemuk consultants said that women

RED MAIDS
Calandrinia ciliata

who were on restricted diets during menstruation or after childbirth ate no meat or fat, but often ate seeds like red maids or chia in place of meat with their acorn mush. The seeds were also eaten by the Salinan (Fages 1937:79) and Miwok (Barrett and Gifford 1933: 152–153).

For the Chumash, perhaps even more important than food was the use of red maids seeds in ritual offerings. For example, Harrington's consultants recounted that when visiting a sacred spring to collect water for curing the sick, a person would scatter offerings of *pil*, chia, shell beads, and tobacco around the edges of the waterhole. They left similar offerings at other kinds of shrines as well, and also placed them in graves.

Archaeological evidence confirms the information Harrington recorded. *Calandrinia* seeds have been found in late prehistoric- and historic-period burials throughout Chumash territory, including the Santa Monica Mountains (King 1969:37), Santa Ynez Valley (Harrington 1928:177–178), Goleta Slough (Yarrow 1879:36–37), and the Channel Islands. One burial contained twelve quarts of these seeds, radiocarbon dated at 600 ± 70 years before present (Orr 1968: 200).

Harrington's consultants said that whenever danger was seen coming, an old woman who had great faith would throw these seeds in that direction three times, saying "hah" each time. This was an offering of sacrifice to *heishup*, the world. Or if men were out in a canoe and the wind caught them, they would throw beads and *pil* into the water as a sacrifice. The missionary explorer Crespí described Indians in the Los Angeles region howling and throwing seeds in the air and on the ground as they met the expedition group (Crespí 1927:148). Though he thought the seeds were meant as a gift, perhaps the intent was quite different.

Neighbors of the Chumash, the Kitanemuk, also used red maids seeds in offerings. At the winter solstice, they arranged red maids seeds, chia, *piñon* nuts (*Pinus monophylla*), and white eagle down in the shape of a cross on a large basketry tray. The ceremonial leader deposited it on a nearby hilltop with prayers and more offerings of seeds and beads. Offerings made at the end of boys' *Datura*-drinking puberty ceremonies also included red maids seeds.

Liliaceae — Lily Family

Calochortus spp. — **MARIPOSA LILY**

(B) nakhaykha'y, kh'a'w, matakh — (Sp) Cacomite

(I) pulash

(V) 'utapits, 'utapikets

Harrington's consultants gave Chumashan names for several kinds of mariposa lilies, often referring to them as "another kind of *cacomite*." The corms or bulbs of *Calochortus* are similar to those of brodiaea, which they called *cacomite*, and were eaten in the same way (see *Dichelostemma*).

The type of mariposa lily which the Barbareño called *nakhaykha'y* (probably *Calochortus catalinae*) had white flowers with black centers and larger bulbs that were very sweet. (This may be the same as Ineseño *pulash*.) A second kind, *matakh* (possibly *C. venustus*), also had white flowers but the bulbs tasted like sweet potatoes. The third kind, *kh'a'w* (possibly *C. luteus* or some forms of *C. venustus*), had yellow flowers. The two Ventureño varieties, as yet unidentified, were distinguished by the size of the bulb: *'utapits* was the size of a small round tomato, and *'utapikets* was smaller, the size of a garlic.

Convolvulaceae — Morning-Glory Family

Calystegia macrostegia (E. GREENE) BRUMMITT subsp. *cyclostegia* (HOUSE) BRUMMITT — **COASTAL MORNING-GLORY**

(B) s'epsu' 'i'ashk'á' — (Sp) Mañana de la Gloria

(I) s'epsu' 'aquqa'w

(V) 'almakhmal 'i suninakhshep

Lucrecia García labeled several specimens of this plant with a delightfully descriptive Barbareño Chumash name that means "Coyote's basket-hat." No information on Chumash uses of any morning glory species was found in Harrington's notes.

Aizoaceae — Fig-Marigold Family

Carpobrotus chilensis (MOLINA) N.E. BR. (Introduced) — **SEA FIG**

(B, I) sto'yots' — (Sp) Tunita Medanal, Jayagüis

(V) shtamhɨl, shtoyho'os

This common "ice plant" probably originated in South Africa (Hickman 1993:128). Although not indigenous to California, it was widespread in coastal dunes at least a century ago and was used by the Chumash. The leaves are succulent, the magenta, many-petaled flowers resemble cactus blossoms, and the fruit is edible—hence the Spanish common name *tunita medanal*, meaning "little cactus fruit of the sand dunes." The Chumash clearly distinguished it from *tunas*, cactus fruit (*Opuntia* spp.)

Harrington's consultants differed on the color of the ripe fruit. Fernando Librado, who had apparently been shown a specimen of *C. chilensis* (Southwest Museum cat. no. 620-G-52B), said the fruit was purple, but Luisa Ygnacio described it as yellow when ripe and about the size of the end of one's thumb. Either way, the Chumash ate these fruits raw and said they were good, very sweet-tasting and just a little salty. This species was also eaten by the Luiseño (Sparkman 1908:232) and by the Southwestern Pomo (Gifford 1967:13). The Chumash and other California Indians may have also eaten the very similar Hottentot fig (*Carpobrotus edulis*).

Scrophulariaceae — Figwort Family

Castilleja spp. — **INDIAN PAINTBRUSH**

(B) swo's 'i'ashk'á' — (no Spanish name known)

(I) mashqupshlét 'akhukha'w

Although no uses were recorded for Indian paintbrush, the Chumash names are colorful: "Coyote's headdress-pin" in Barbareño, "Coyote's rectum" in Ineseño.

Scrophulariaceae — Figwort Family

Castilleja spp. — **OWL'S CLOVER**

(B) ste'leq' 'ipistuk' — (no Spanish name known)

(I) stelek 'ipistúk

Lucrecia García labeled several specimens of owl's clover (formerly *Orthocarpus* spp.) with a Barbareño common name meaning "ground squirrel's tail." Neither she nor other Chumash consultants mentioned any uses for this plant. Pomo people made dance wreaths of the flowers for the spring strawberry festival; their name for this plant means "fly eyes" (Goodrich, Lawson, and Lawson 1980:35).

Rhamnaceae — Buckthorn Family

Ceanothus megacarpus NUTT. — **BIGPOD CEANOTHUS**

(B, I, V) sekh — (Sp) Palo Seje

The Chumash distinguished between white-flowered ceanothus, which they called *sekh*, and blue-flowered forms, *washiko*. More than one species was included in each of these categories, but plant specimens that Harrington's consultants labeled *sekh* were all *Ceanothus megacarpus*. The two kinds were generally considered to have different properties. Although wedges for splitting canoe planks could be made of either *washiko* or *sekh*, most other uses reported for ceanothus referred to *washiko*, discussed below.

Rhamnaceae — Buckthorn Family

Ceanothus oliganthus NUTT. — **HAIRY CEANOTHUS**

C. spinosus TORREY & A. GRAY — **GREENBARK CEANOTHUS**

(B, I, V) washiko — (Sp) Palo Colorado

Chumash speakers recognized two principal categories of ceanothus: white-flowered (*sekh*) and blue-flowered (*washiko*). Both *Ceanothus spinosus* and *C. oliganthus* were *washiko*, which had a greater variety of uses than *sekh*. The Spanish name *palo colorado* may refer particularly to *C. spinosus*, which is sometimes known as "red heart" in English.

Wood of *washiko* lasted for a long time in the ground, making it valuable in building fences and corrals, and also for offertory poles at shrines. A feathered pole made of *palo colorado* marked the Depository of the Things of the Dead, a shrine near Ventura. Another shrine was described as being a circle of similar poles fifteen feet across, each pole about two inches thick and four feet high, unpainted but decorated with feathers.

Washiko was commonly used to make the digging stick, a very efficient tool for digging roots and bulbs. The stick was about four feet long and a bit less than two inches in diameter. The Chumash shaped and hardened the end by charring it in fire, scraping it to a point, and then placing it alternately in fire, water, fire again, and so on. They

GREENBARK CEANOTHUS
Ceanothus spinosus

often weighted the digging stick with a pierced, doughnut-shaped stone, twisting it onto the shaft about a foot up from the bottom, they said, depending on the depth one wished to dig.

Chumash people also made *washiko* awls for basketmaking and pry bars for collecting abalones. Wedges for splitting canoe planks could be made of either *washiko* or *sekh* types of ceanothus.

The Chumash, like several other California peoples, made rattles from the cocoons of ceanothus silk moths (*Hyalophora euryalus*), which they said could be found in *palo colorado* thickets. Two or more of these cocoons, with small stones placed inside, were pierced or tied onto a slender stick handle.

Gentianaceae — Gentian Family

Centaurium venustum (A. GRAY) ROBINSON — **CANCHALAGUA**

(no Chumash name known) — (Sp) Canchalagua, Conchalagua

Canchalagua was boiled into a tea drunk as a tonic and blood purifier (Birabent n.d.; Bard 1894:40). Harrington's Chumash consultants were aware of the medicinal use of this plant, but they were divided on whether it was a remedy of the old-time Indians or of the whites. Other southern California Indian groups also used *canchalagua* as a medicine (Palmer 1878:652; Blochman 1894b:40; Sparkman 1908: 230), as did the Spanish Californians (Saunders 1920:207).

Rosaceae — Rose Family

Cercocarpus betuloides TORREY & A. GRAY — **MOUNTAIN-MAHOGANY**

(B) pɨch — (Sp) Palo Fierro

The common name "ironwood," or *palo fierro* in Spanish, is applied to many different plants that have dense, hard wood. The Chumash sometimes made digging sticks of mountain-mahogany. Juan de Jesús Justo said this wood grew only in the mountains; it was strong and very hard and was used especially for digging brodiaea bulbs (*Dichelostemma*). Henshaw recorded the name as "burtch" in an attempt to write an English speaker's pronunciation of the Chumashan name (Henshaw 1887:7).

Euphorbiaceae — Spurge Family

Chamaesyce spp. — **PROSTRATE SPURGE**

Euphorbia spp. — **SPURGE**

(no Chumash name known) — (Sp) Golondrina

A nineteenth-century Ventura physician stated that "*golondrina, Euphorbia maculata*," was used by the local Indians for skin diseases, warts, and cataracts (Bard 1894:6; echoed by Benefield 1951:22). This species (now *Chamaesyce maculata* L., called spotted spurge in English) was introduced to California from the eastern United States. *Golondrina*, or *golondrinera*, was said to be used for fever by Indians and settlers in the Santa Barbara area (Birabent n.d.).

Harrington's consultants mentioned no uses for *golondrina*, and its identification is somewhat doubtful. Lucrecia García labeled specimens of spineflower (*Chorizanthe*) as *golondrina*, but most other references pertaining to California and the Southwest indicate that *golondrina* was generally a *Euphorbia* (e.g., Ford 1975:201–202). In the Southwest, prostrate *Euphorbia* spp. were known as *yerba de la golondrina* and used as poultices, either fresh or dried and steeped in wine, for bites of rattlesnakes and tarantulas (Rothrock 1878:50).

Here I must repeat the caution against experimentation with any of the remedies mentioned in this book. Most *Euphorbias* and many of their close relatives have very caustic juice that may cause blisters and should never be put into the eyes or taken internally.

Asteraceae — Sunflower Family

Chamomilla suaveolens (PURSH.) RYDB. (Introduced) — **PINEAPPLE WEED**

(no Chumash name known) — (Sp) Manzanilla

Formerly known as *Matricaria matricarioides*, this plant is a common weed which some authors have considered to be native to the Chumash area (Smith 1976:302). According to María Solares, however, *manzanilla* did not occur in Santa Ynez in the early days. Rosario Cooper indicated that it was one of the medicinal plants gathered by the Obispeño Chumash but did not specify how it was used (Greenwood 1972:83). No native names were reported.

Early authors report that *manzanilla* tea was taken for pains in the stomach and nerves (Birabent n.d.); for gastrointestinal disorders in general (Blochman 1894b:40); to induce perspiration and to reduce pain (Bard 1894:6); and for cramps and colic, and to suppress menstruation (Benefield 1951:22). *Manzanilla* tea was also reportedly used for the liver, in childbirth, and to treat catarrh and dysentery (Garriga 1978). More recently, Chumash descendants and others have been known to drink this tea as a beverage or to treat stomachache, indigestion, or *empacho*, and the plant is sometimes applied as a poultice to reduce inflammations (Gardner 1965:297; Weyrauch 1982:11).

The commercially available *manzanilla* in common use today is true chamomile, *Chamaemelum nobile* (formerly *Anthemis nobilis*), a garden plant introduced from Europe as a medicinal. *Chamomilla* and *Chamaemelum* are closely related and have similar, scented foliage; both are widely used for many medicinal purposes among Spanish-speaking peoples in the Southwest and Mexico (Ford 1975:232–233). Among California Indians, the Pomo employed a decoction of *Chamomilla* leaves and flowers to check diarrhea (Mead 2003:256). The Diegueño, Maidu, and Coast Miwok drank the tea to reduce fevers (Strike 1994:91). The Ohlone made a salve of the seeds for infected sores, as well as using the leaf decoction for stomach trouble (Bocek 1984:255). It is possible that medicinal use of *Chamomilla* proliferated in California after *Chamaemelum* and its uses became familiar to New World peoples. Though widely used, chamomile is listed as toxic and may induce anaphylactic shock (Hickman 1993:226).

Chenopodiaceae — Goosefoot Family

Chenopodium berlandieri MOQ. — **PITSEED GOOSEFOOT**

(B) we'lel — (Sp) Choale, Chuale

(I, V) welel

Harrington's consultants from Santa Ynez and Ventura said that Chumash people used to eat the seeds of this plant and also ate the young plants and leaves as greens.

Chenopodiaceae — Goosefoot Family

Chenopodium californicum (S. WATSON) S. WATSON — **SOAP PLANT**

(B) su'núk' — (Sp) Raíz de Lavar

(I) 'akhwayɨsh

(V) choch

This plant might be confused with another "soap plant," the bulb-forming *Chlorogalum pomeridianum*, because both have the same common name in English. Chumash speakers, however, carefully distinguished between these two unrelated lather-producing plants. The Spanish common names also differ.

Chenopodium roots are long, slender and hard. They were dug with a digging stick and scraped or grated for use in washing the hair, skin, or clothing. The Spanish Californians considered this plant superior to *Chlorogalum* for washing wool and made much use of it (Harrington 1986; Blochman 1894a:10).

To prepare a medicinal tea for the treatment of consumption, the Chumash scraped the bark from *Chenopodium* root and boiled it in water. The patient drank the strained liquid, which acted as an emetic and expectorant. This treatment was followed by drinking tea of *yerba mansa (Anemopsis californica)* for three days.

Liliaceae | Lily Family

Chlorogalum pomeridianum (DC.) KUNTH

SOAP PLANT, SOAPROOT

(B) qi'w

(Sp) Amole

(I) kot'; chnuy (fiber)

(O) tqupa'

(V) pash

The Chumash found several uses for this common plant of the lily family. They dug the bulbs with a digging stick in the springtime and carried them back to the village for processing. The outer brown fibers were removed and saved for separate use.

The fleshy core of the bulb was crushed and mixed with water to create suds for washing clothing and hair. Clothing not rinsed thoroughly after such a cleansing caused skin irritation. With their fingers, women applied the juice from soaproot bulbs to their bangs to make them lie flat against their foreheads; it also made their hair glossy.

The Chumash used soaproot in preparing hides other than deer skin. To work a whole fox skin, for example, they turned the skin inside out and washed it with this soap to cut the grease and soften the hide. They pounded the soaproot bulbs and rubbed the resulting paste on the flesh side of the skin until it was covered with suds. Then they turned the hide right side out and worked it with the hands so that it would dry in its natural shape.

As the principal, if not the only, fish "poison" used by the Chumash, soaproot did not affect the edibility of the fish. The bulbs were crushed and stirred into quiet pools in freshwater streams. Saponins, soapy chemicals in the plant, apparently coated or paralyzed the gills, stupefying the fish so they floated to the surface. People would then enter the water and pick up the fish with their hands. Soaproot fishing continued well into the twentieth century in the Santa Ynez River and was also practiced by the Barbareño and Ventureño Chumash.

The Chumash made brushes from the coarse, brown fibers surrounding the soaproot bulb. They cleaned the fibers by pounding them on a rock, and then separated them by combing. Small bundles of this fiber were tied together at the upper ends with Indian-hemp string to make a handle, which they covered with tar—not the glue from boiled soaproot bulbs as was the practice among the Miwok and some other California peoples (Barrett and Gifford 1933:215). The base ends of the fibers formed the working end of the brush, which they trimmed to an even length and singed. Some Chumash women made brushes decorated with beads for sale, and rich Spanish Californians would pay twelve *reales* for them.

These brushes were very handy for cleaning flour off a metate, or to clean the metate after grinding. Sometimes people might use an old hairbrush for this purpose, but they never used the same brush for their hair and for cooking. Bulbs with particularly stiff fibers were selected for making hairbrushes; some bulbs had softer fibers and were not as useful for this purpose. Although Chumash men seldom brushed their hair, women brushed theirs frequently by holding it out with one hand and brushing along it with the soaproot brush. When a woman was pregnant, she would only scratch herself with a soaproot brush. It was believed that her skin was so tender that her fingernails would leave permanent scratches. Some non-Indians employed the entire soaproot bulb, with the fibers still attached, as a scrub brush for laundry and housecleaning (Blochman 1894a:10).

Some Chumash shamans made large images of bears by covering wooden frameworks with brown soaproot fiber. The head, eyes, snout, and legs of the life-sized figure were all like a bear's. The shaman would sit inside it and ambush people. He wore a fine mesh net of soaproot fiber to protect him from the arrows of his victims. If a bear could not be killed by arrows, the Chumash believed it was because the "bear" was really a shaman wearing this soaproot fiber armor.

Other California peoples roasted soaproot bulbs with the husks still on in underground earth ovens, but the Chumash do not seem to have done so. They did eat the young shoots of soaproot after roasting them in ashes, saying they tasted like yucca, but hot like pepper. This plant was widely used throughout California for food, fish poison, soap, glue, brushes, and other purposes (Heizer and Elsasser 1980: 244).

Polygonaceae — Buckwheat Family

Chorizanthe spp. — **SPINEFLOWER**

(no Chumash name known) — (Sp) Golondrina

In early documents, *golondrina* is identified as *Euphorbia maculata* (now *Chamaesyce maculata*), an introduced weed from the eastern U.S.

However, Lucrecia García labeled two specimens of *Chorizanthe* as "*golondrina*." It is uncertain whether both of these plants were known as *golondrina* in this region, and whether their use was indigenous to the Chumash or introduced historically. *Golondrina* was used for fever, skin diseases, and warts (Bard 1894:6; Birabent n.d.; Benefield 1951:22).

Asteraceae — Sunflower Family

Cirsium spp. — **THISTLE**

Hypochaeris radicata L. (Introduced) — **HAIRY CAT'S EAR**

Sonchus oleraceus L. (Introduced) — **COMMON SOW-THISTLE**

(B, I) qayɨsh — (Sp) Cardo

(V) ts'aqsmi

Lucrecia García collected specimens of these three thistle-like plants and labeled them "*khayish*." Some members of the genus *Cirsium* are native to California, but the other two specimens are weeds introduced from Europe.

The Chumash may have used several kinds of thistles as food. One of Harrington's consultants specifically mentioned eating the tender leaves of *qayɨsh*, a large thistle with a yellow flower. Others described this plant as the common dandelion-like weed one or two feet tall with spines, a yellow flower, and white juice. Dandelion (*Taraxacum officinale*) is also an introduced weed.

The Ohlone ate the stems of *Cirsium* either raw or boiled, and they also made various medicinal uses of the root and stems (Bocek 1984:254). The Cahuilla ate the bud at the base of the plant (Bean and Saubel 1972:55)

Portulacaceae — Purslane Family

Claytonia perfoliata WILLD. — **MINER'S LETTUCE**

(B) shilik' — (Sp) Petota

(I) shilik

Seeds of miner's lettuce, which closely resemble those of red maids (*Calandrinia*), were a traditional food of the Chumash. Surprisingly, none of Harrington's consultants actually described eating fresh raw leaves of miner's lettuce, though many other California peoples did and continue to do so (Bocek 1984:251). They did say that the leaves were boiled and eaten. This practice may have been introduced

MINER'S LETTUCE
Claytonia perfoliata

in historic times, however, for there is little evidence that the Chumash formerly ate cooked greens.

The Spanish term *quelite* is a general term for greens, especially those cooked for eating. The equivalent Chumash words are (V) *chomoy* or (B) *shomoy* and (B, I) 'anɨtɨsh. Plants mentioned as falling into this "greens" category included introduced sweet-clover (*Melilotus* spp.) and bur-clover (*Medicago polymorpha*), as well as native clovers.

CLEMATIS
Clematis sp.

Ranunculaceae — Buttercup Family

Clematis lasiantha NUTT. — **CHAPARRAL CLEMATIS**
C. ligusticifolia NUTT. — **CREEK CLEMATIS**
(B, P) makhsik' — (Sp) Barba de Chivo
(I) 'alamakhwak'ay
(V) makhsik

These vines with cream-colored flowers climb over shrubs and trees in streamside woodland and chaparral. They have plumed seeds borne in tufts, hence the Spanish name *barba de chivo*, "goat's beard." Clematis was used medicinally. Chumash people rubbed the leaves on their skin to treat ringworm, skin eruptions, and perhaps venereal disease.

Spanish Californians made an infusion of *Clematis ligusticifolia* leaves to treat sores and cuts on horses, and some Indian peoples used it to treat colds and sore throats (Jepson 1925:392; Munz 1959:103). The Ohlone applied clematis compresses for chest pain (Bocek 1984:248).

Cornacea Dogwood Family

Cornus glabrata BENTH. **BROWN DOGWOOD**

C. sericea L. subsp. *occidentalis* (TORREY & A. GRAY) FOSB. **CREEK DOGWOOD**

(B) wiliq'ap' (Sp) Iriris

(I) wiqap'

(V) wiliqap

Harrington's Chumash consultants spoke of a plant they called *iriris*, which was used to make fishing poles, canoe ribs, bows and arrows, and cradles. Although Yridisis Creek in the San Julian area of western Santa Barbara County is said to have been named for *Cornus glabrata*, the Spanish common name *iriris* most likely referred to creek dogwood (*Cornus sericea*, formerly *C. stolonifera*), a sturdier and more widespread species.

Fishing poles were made with two sections of dogwood stems, each about four feet long and an inch in diameter, by overlapping the pieces a few inches and wrapping them with cord. The fishing line was only about two feet long, but the pole was flexible and springy. The Chumash would catch bullhead (probably *Cottus* sp.) by sticking the pole under the rocks and pulling it out. When finished, the fisherman could take apart the two sections of the pole for ease in carrying.

A popular game was hoop-and-pole, played by rolling a four-inch-diameter hoop along the ground and attempting to throw a pole through it as it moved. The poles for this game were sometimes made of dogwood.

Harrington's consultants mentioned the use of dogwood for bows and arrows. Local folklore also says the Chumash made arrows of *Cornus glabrata* (Clifton Smith, personal communication 1980).

Chumash canoes originally did not have any interior framework, but in the historic period canoes were sometimes made with ribs in imitation of European boats. They had three or four ribs, put in place after the canoe was finished. To form the ribs, the canoe maker cut poles of dogwood six to eight feet long and an inch and a half in diameter, then peeled and split them lengthwise. He bent them to fit the hull, notching the inside of each rib to facilitate bending, and placing it with the flat face resting against the inside of the planks. To attach the ribs, he could either drill holes with an auger and insert small nails, or stake and tie the ribs in place.

To guide the fitting of the planks and achieve the desired shape of the canoe, the craftsman used a measuring device of dogwood as a template. He heated the dogwood stick in glowing coals and, while it was still hot, bent it to the right shape and tied the ends together. If the stick cooled prematurely, it would break or not bend well. This measuring bow was left in the canoe throughout construction and taken out when the boat was finished (Hudson, Timbrook, and Rempe 1978:44, 75)

Chumash baby cradles had crosspieces of dogwood lashed to the forked willow framework or inserted into holes drilled in it. Six more dogwood sticks were tied on, running the length of the cradle. People liked to use dogwood for this, they said, because it grows straight and does not require heating to be straightened.

Euphorbiaceae — Spurge Family

Croton californicus MUELL. ARG. — **CALIFORNIA CROTON**

(B) 'ilaq — (Sp) Barbazo, Barbasco

(V) smakhna'atl

See also *Atriplex*. Both genera seem to have been in the same Chumash folk category and it is difficult to distinguish them on the basis of Harrington's notes and plant specimens.

Apparently croton was one of the plants burned in the mission era to obtain ashes for use as lye in soapmaking. Medicinal uses are also attributed to this species, a tea of "*barbaza*" to be drunk hot at bedtime for colds (Birabent n.d.). A similar medicinal use is described for croton among the Cahuilla (Bean and Saubel 1972:56). Fernando Librado said that *barbasco*, called *smakhna'atl* in Ventureño, was made into a salve and rubbed on the skin for rheumatism. This use is reported for croton elsewhere in the Hispanic Southwest (Blochman 1894c: 162–163; Ford 1975:90, 140–141).

Cucurbitaceae Gourd Family

Cucurbita foetidissima KUNTH **WILD GOURD**

(B, I) mo'kh (Sp) Calabazilla, Chili Coyote, Chilecayote

(V) mo'okh

The Chumash sliced and scraped the root of this ill-smelling vine to use as soap. It was very strong and cleaned effectively, but it was irritating. If clothing washed in this were not rinsed well, the unfortunate wearer could break out in leprosy-like sores or a swelling on the neck. Some people, particularly Spanish Californian women, also used the fruit of this gourd like lye in washing. Many Spanish laundresses in the late nineteenth century still employed the root (Blochman 1894a:10). Chumash people seem to have utilized wild gourd less commonly than other soaps.

Wild gourd also had medicinal uses. The Chumash crushed the tendrils, put them in water, and drank this as a strong, bitter purgative, or boiled and pounded the roots in water for the same purpose. A treatment reported for rheumatism involved rubbing the body with *chuchupate (Lomatium californicum)* and then drinking this gourd-root decoction for five days. An instance of nosebleed caused by sorcery was said to have been cured by having the victim snuff a strained mixture of raw gourd root and water up his nose.

The Chumash made containers, drinking cups, and dippers out of wood, basketry, and gourds. It is uncertain whether the practice of using gourds in this way was indigenous or began in mission times with the introduction of cultivated bottle gourds, *Lagenaria leucantha.*

Gourd rattles were an important part of the bear disguise used by certain Chumash shamans (see *Chlorogalum* for a description of this outfit). The shaman hollowed out two wild gourds and wore them on a string around his neck, one at each end, hanging down in front of his chest. When the "bear" ran, these gourds sounded like the snorting of a bear at every step.

Luisa Ygnacio told Harrington that there were many wild gourd plants growing back of the Thomas Hope ranch house and at La Goleta. She remembered one time when several Indians were at *Helo'* village, near the Goleta Slough, cutting willow to put under the thatching for the roof of the Cieneguitas chapel. The men played by throwing *mokh* gourd fruit at each other until they were soaked.

Seeds of this species were reportedly eaten by the Luiseño and Cahuilla (Sparkman 1908:229; Bean and Saubel 1972:57–58). Use of the root, fruit, and leaves as cleanser and medicine was widespread.

Cyperaceae — Sedge Family

Cyperus esculentus L. — **CHUFA, YELLOW NUT-GRASS**

(no Chumash names known) — (no Spanish name known)

Several hundred charred nut-grass tubers, presumably food remains, were found in two pit-ovens during archaeological excavations at San Antonio Terrace, north of Lompoc (Miksicek 1998). This plant has not been reported from other sites in the Chumash region, nor is there any mention of it in Harrington's notes. Perhaps it was used in only a limited area or at a particular time period, and the Chumash people he consulted were not familiar with it. This obscure tidbit of archaeological evidence is a good reminder that, even when extensive ethnographic, historical, and archaeological information are all pieced together, our understanding of Chumash plant knowledge will never be complete.

Nut-grass is widespread in moist areas throughout California. It was a significant food source for the Owens Valley Paiute, who are thought to have cultivated it on a large scale (Lawton et al. 1976), and the starchy tubers were also eaten by the Pomo (Goodrich, Lawson, and Lawson 1980).

Datiscaceae — Datisca Family

Datisca glomerata (C. PRESL) BAILLON — **DURANGO ROOT**

(B) 'ansiwa'wu'y — (Sp) Yerba Colorada, Raíz Colorada

(V) 'aluqchahay 'isakhpilil

The Spanish and Ventureño names refer to the reddish color of the root. This is one of the few dyestuffs reported for Chumash basketry; Harrington's consultants said this root was one of two they used for dyeing *Juncus* yellow. See also *Rumex hymenosepalus.*

Durango root was used medicinally for sore throat. The fresh root could be chewed raw or the juice pressed out of it and swallowed. The Chumash often collected the root and dried it for later use as a tea or gargle to treat sore throats (Harrington 1986; Birabent n.d.). The Ohlone also utilized this plant as a dye and sore throat medicine (Bocek 1984:250).

Solanaceae — Nightshade Family

Datura wrightii REGEL — **JIMSONWEED, TOLOACHE**

(B, Cr) mo'moy — (Sp) Toloache

(I, V) momoy

Most published literature refers to *Datura meteloides* or *D. inoxia*, but these names have been superseded by *D. wrightii* (Hickman 1993:1070). This is the most common *Datura* in the Chumash area, although an introduced species native to Mexico, *D. stramonium*, also occurs here.

DATURA
Datura wrightii

Important Note: All parts of this plant are highly poisonous. Every year people die from ingesting it through misidentification or foolhardiness. The dangerous compounds can also be absorbed through the skin.

Despite *Datura*'s extreme toxicity, the Chumash and other peoples highly valued this plant for its vision-inducing and pain-killing properties. It was probably the single most important medicinal plant of the Chumash, since in their culture medicine was an integral part of religious beliefs and practices.

The purposes and methods of *Datura* use by the Chumash have been described in detail by Applegate (1975a). In general, *toloache* was taken for three principal purposes: to establish contact with a supernatural guardian who would provide protection, special skill, and a personal talisman; for clairvoyance, such as contacting the dead, finding lost objects, seeing the future, or seeing the true nature of people; and to cure the effects of injury, evil omens, or breaches of taboo, and to obtain immunity from danger (for example, a person whose form was assumed by a coyote was in grave danger and should take *toloache* right away to prevent soul loss and death).

Depending on the purpose for which it was to be used, *toloache* was most often ingested as a drink made from the roots or leaves. Only small portions of the root of each plant were collected, so as not to damage the plant. Four or more roots were well pounded in a small or medium-sized mortar, then put into a container and soaked in cold water. The mixture was stirred, allowed to settle, and strained into another vessel. The effects of *Datura* vary with the age and size of the plant, the soil in which it is growing, the season in which it is collected, the strength of the brew, and the condition of the person taking it. The Chumash were quite aware that it was potentially lethal, and they took precautions not to offend the *toloache* spirit as well as to gauge the correct dosage. A strong man might be given a measure equal to the first joint of the forefinger. A weak person would be given half as much.

The first time a person took *toloache* was an initiation ritual in the sense that it prepared him or her for adult life by providing a spirit guardian, or "dream helper," to give guidance in the future and help the person to become successful in life. A young boy or girl at the age of puberty was given toloache individually in the house of the specialist who prepared the potion. After drinking it, they would fall into a coma for a period of time. When they regained consciousness the specialist asked them what they had seen and interpreted their visions. In the dream, they might have seen a spirit or force that spoke to them in a religious language and gave them an *'atishwɨn*, or talisman symbolic of its protection, to wear around the neck or waist. They could call upon this spirit guardian in times of crisis.

The dream helper was felt to be very reliable. For example, when Palatino took *toloache* he saw a *khelekh*, Peregrine Falcon; and because he had seen this bird in his *toloache* dream, he saw it again later when he was in distress. He and others were out in three canoes, fishing far offshore, when they were caught by the wind. Two canoes sank in the huge waves, and three men were drowned. The third canoe was also in danger. Palatino stood up in the canoe and called on his spirit helper three times. From the sky came the sound of the Falcon. Palatino saw it and followed its flight eastward. The bird led them safely to shore near the mouth of the Santa Clara River (Hudson, Timbrook, and Rempe 1978:159).

Some people failed to have a vision the first time they took *toloache*, so they might take it again. It was considered better to have a dream helper, but not everyone did. Most people took *toloache* once; some, several times. Some parents encouraged their children to take *toloache* to increase their powers of curing or bringing good fortune to the family.

After puberty, both men and women could take *toloache* at any time. This was done on an individual basis and usually at home. Establishment of contact with the supernatural might be sought "in order to become strong, in order not to fear anyone, to prevent snakes from biting them, and that darts and arrows may not pierce their bodies" (Geiger and Meighan 1976:48). In the *toloache* vision, one might make prayers for particular benefits or skills: success in hunting, protection against bears, power in games—Winai prayed for success in billiards—or ability in horsemanship. Some individuals drank *toloache* repeatedly through a desire for even greater supernatural power or invincibility. *Toloache* drinking was regarded as a dangerous practice, however, and those who did it to excess risked mental damage or even death.

Doctors and other kinds of shamans used *toloache* to strengthen their connections to the spirit world, and they often had several dream helpers to aid them. Shamans would take *toloache* and observe certain restrictions before handling ritual objects, such as charmstones. One account describes a shaman drinking a quantity of the *toloache* decoction and then biting the charmstone, which enabled him to fly rapidly wherever he wished to go (Yates n.d.). The feeling of flying is one of the effects of *Datura* spp. that made the plant an important ingredient in early European witches' brews (Armstrong 1986:39–40). Physiological aspects of *Datura* ingestion have been summarized by Baker (1994).

Harrington's consultants gave examples of *toloache* being used to counteract the effects of evil omens and breaches of taboo:

> Narciso, a Ventureño, worked as a cook for the Noriegas in Santa Barbara, and his wife Martina lived in La Patera. Once Narciso saw Martina climbing the stairs, but it turned out to be a coyote that had assumed the form of Martina. Narciso was very sad and went to the real Martina and told her what he had seen. Martina said, "It is necessary to drink *toloache*." So they both drank *toloache*, and nothing bad happened to them (Blackburn 1975:295).

> Once an islander who was staying at Santa Inés was coming home late at night. He saw a dog chewing on a beef head that had been thrown out, and he threw a rock at it. The dog spoke to him and said, "Why do you throw rocks at me? If someone threw rocks at you, you wouldn't like it!" The man came home crying and told his wife what had happened. They both left for the man's mother's home at San Marcos, where she administered *toloache* to both of them. In his dream, the man saw the dog and it said it would do nothing to harm him. When he woke, he knew everything was all right. That dog died at daybreak (Blackburn 1975:296).

> An American man once went to Santa Cruz Island with a Ventureño boy. While they were gathering abalone, the American found a cave in the rocks and looked into it. He saw two old men inside—one had a bullroarer and the other had an elderwood flute—and they were dancing and singing as do the *'antap* in the ceremonial enclosure. When the water dashed over them they were unaffected by it, it was as nothing to them. On the way back from the island, the American was drowned and the Chumash boy returned to Ventura alone. He told his mother what he had seen and she immediately gave him *toloache* to counter the effects (Blackburn 1975:290–291).

These accounts were given about seeing the future or the true nature of people:

> Paula was given toloache to help her husband, Pedro Celestino, to recover from a beating. When she woke, she said she had seen Pedro kill his wife. Iluminado, the toloache master, said that if she wanted to save her own life, she must take toloache again. This time she saw nothing. Iluminado told her she was safe, but must never disobey her husband. Paula lived for some time after that but died from disease. Pedro Celestino then married a second wife, Juana. When he was working as a cook for Anastacio Carrillo in Santa Barbara, he slashed Juana with a knife and killed her. That was what Paula had seen in her toloache dream (Hudson 1979:69–70).
>
> When Winai took toloache after breaking an arm and a leg, in his dream he saw Cholo Roberto standing in the house with his arm unnaturally long. He was placing *'ayip* [a powerful, dangerous substance] in the upper corner of the room. [Cholo Roberto's intent was to cause harm to the people who lived there, and the toloache enabled Winai to see this.] (Hudson 1979:71–72)

Toloache made a difference in what happened to a person after death as well as during life. Some Chumash believed that after death the soul journeyed to *Shimilaqsha*, the Land of the Dead. It traveled westward, stopping at Point Conception before undergoing a number of trials. As the soul crossed a pole bridge over a body of water, two giant monsters rose up out of the water, trying to frighten it. If the soul was of a person who had not drunk *toloache*, who had no dream helper and did not know about the old religion, it fell into the water and the lower part of its body turned into that of a frog, turtle, snake, or fish. The old people said that someone who drank *toloache* always passed the pole safely, for they were strong of spirit (Blackburn 1975:100).

María Solares told Harrington that the people who lived at La Quemada, near Point Conception, would sometimes see the soul of a person who was dying at another village. It was possible that the soul might be convinced not to go to the Land of the Dead, so anyone at La Quemada who recognized the soul would hurry to the village where that person lived. If the person whose soul it was would then drink a lot of *toloache*, there was a chance that the person would recover and not die.

In addition to being taken to restore spiritual balance resulting from evil omens or breaches of taboo, *Datura* seems to have been taken as a painkiller for serious injuries and broken bones. It had other medicinal uses as well. Harrington's consultants said that the four principal Chumash medical treatments were drinking sea water, taking *toloache*, eating red ants, and cauterizing (see *Artemisia douglasiana*). These were considered good for just about anything, particularly as remedies of last resort when all else failed. For ridding the body of tapeworms, *toloache* was said to be next in efficiency after sea-water drinking. Crushed *Datura* leaves were applied as a poultice for hemorrhoids (Birabent n.d.) or put on the skin over any aching place. A poultice for wounds was made of roasted *Datura* roots, which were held in place with splints of elderberry wood (Hudson and Blackburn 1986:295).

When *toloache* was to be administered to a sick or injured person, the parents summoned all the other relatives. They gathered in the afternoon and danced, shouted, and stayed around to keep the sick person from sleeping until the medicine was given in the early morning. Care was taken to give a small dosage, because some people died from taking too much. Then the patient sat down and held a stick vertically in both hands. After a while he would begin to tremble, and then he would fall over. At this time, broken bones could be set or other necessary treatment performed.

One man who drank *toloache* after he had broken an arm and a leg said that right after taking it he felt a great heat all through his body followed by a terrible itching. Observers described him as seeing visions and talking to himself like a crazy man. He even got up, completely oblivious to the pain of his broken leg. After about fifteen minutes he fell sound asleep. When he awakened, the pain had abated.

When the patient awakened, he was not allowed to drink water for two days, though he might rinse his mouth with a little warm water. He was permitted to eat only light gruel and clams, rabbit, or other light meat, and if married must sleep alone for twenty-one days. After that he could take *toloache* again.

Whether *Datura* was taken to induce visions, to restore harmony, for clairvoyance, or for other spiritual or medicinal purposes, restrictions such as these were an integral part of the use of this plant. Most commonly mentioned were abstention from sexual intercourse, both before and after drinking *toloache*, and various degrees of dietary restrictions. The most extreme was a total fast, in which the only thing to be eaten was tobacco mixed with lime (see *Nicotiana*). In most cases it was regarded as important to refrain at least from eating meat or grease, and somewhat less important to abstain from salt or sweets.

The more power a person sought in *toloache*, the more stringent the restrictions and the longer they must be followed.

When *Datura* was to be taken as a medicine, fewer and less stringent restrictions applied. In cases of emergency, the drink was prepared by a relative of the patient, especially the mother or grandmother. Since treatment was urgent, the patient would not have time to observe the necessary sexual and dietary restrictions before taking *toloache*, but would be required to do so afterwards.

There were no seasonal restrictions on *Datura* use, although the Chumash name for January as the "month of *toloache*" may indicate there was formerly a practice of drinking *Datura* in the early spring, as the Yokuts did (Applegate 1975a:8). One Chumash myth states that in the old times, when animals were people, Coyote used to give *toloache* every year to all the eight-year-old boys in the village (Blackburn 1975:196). This may also indicate that a particular time of year was regarded as appropriate for taking *Datura*.

In Chumash astrological beliefs, persons born in January, the "month of *toloache*," were likely to overestimate their own worth and to be filled with self-love. Fernando Librado told Harrington, "When a man knows how to use virtues properly he succeeds. *Toloache* has virtues also, but it must be properly used" (Blackburn 1975: 101).

Datura was thought to be a constituent of rattlesnake venom. María Solares said that a rattlesnake never bites anybody without planning it in advance. When it plans to bite a certain person, it goes and sinks its fangs into a *toloache* root. When the snake is well filled with *toloache*, it is then ready to bite, and the person it bites will die quickly.

The Chumash name for *Datura*, *momoy*, refers both to the plant and to one of the First People, who was transformed into the *Datura* plant at the time of the mythical flood. In Chumash oral tradition, *Momoy* is portrayed as a very wealthy old widow who lives in a remote place by herself or with a daughter. In the stories she often acts as a fostering grandmother or adoptive mother. While she does not directly affect the outcome of events or overcome malevolent beings, she can foresee the future and warn others of the consequences of their actions. By drinking water in which she has washed her hands, other characters in the myths can partially share this ability. They have visions which seem to portend the future, or which provide access to sources of supernatural powers. This personification of *Datura* in myths suggests that the Chumash had a long and extensive involvement with the plant. It does not appear to such a degree in the mythology of any

OLD WOMAN MOMOY

Momoy (Datura), *an old woman with the power to foresee the future, is the only plant that appears as a person in Chumash myths.*

other California Indian group (Blackburn 1975:36). Many animals appear as people in Chumash myth, but *Datura* is the only plant to be viewed in that way.

Harrington's Kitanemuk consultants said that the "*toloache* religion" was of the coastal people, i.e., the Ventureño Chumash. But they did provide information about use of *Datura* in their own initiations which differs somewhat from the Chumash practice. Among the Kitanemuk, *toloache* was not administered individually in the specialist's house but rather in the dwelling house to groups of boys at the age of puberty to help them make contact with a spirit guide, usually an animal. They remained unconscious until the next afternoon, and during their visions each one received something as a talisman. He kept this in his house and did not wear it as an amulet. The headman of the ceremony remembered what each boy had seen, and went up into the hills before sunrise the next day to work for the boys. He prayed and left offerings of down, seeds, tobacco, and beads, after which the boy was a man. The Kitanemuk said that *toloache* would be very bad for girls. They were not given any such initiation ceremony although they did follow certain restrictions at their first menstruation.

The Kitanemuk said that their people took *toloache* only in winter, in cold weather, as it would be too hard on a person later in the year. They discussed this seasonal restriction in connection with taking *toloache* in adulthood, for example as a cure. An old woman or old man dug the root, saying prayers to the plant. The roots were roasted in hot ashes, pounded, and mixed with cold water. The drink was administered at dusk and the patient remained comatose for three days. When he awakened, he was still hallucinating and talking to things; other people did not stay with him at this time but waited outside until he was himself again. For a month afterward, he did not bathe or eat meat, and he spent much time in the hills alone. When people took *toloache* for medicinal reasons, they did not expect to make contact with the spirit world.

Datura species have been widely employed for their hallucinogenic and medicinal properties. In the Southwest and northern Mexico, many peoples used *Datura* medicinally to treat a variety of conditions, including wounds, inflammations and sprains, hemorrhoids, boils, lice, asthma, and headache (Ford 1975:312–313).

The botanist Willis Jepson believed that *D. meteloides* (now *D. wrightii*) originated in northern Mexico but was introduced to California before American colonization, and possibly in pre-Hispanic times, because of its importance to indigenous peoples (Jepson 1943:457). More recently, it has been suggested that the plant was introduced by the early Spanish themselves (Hickman 1993:1070), but that seems extremely unlikely. It is my opinion that, in view of the complexity of well-entrenched cultural traditions surrounding *Datura*, it must have arrived here long before the first Spanish visitors in the sixteenth century.

Apiaceae — Carrot Family

Daucus pusillus MICHAUX — **RATTLESNAKE WEED**

(B) s'akhiyɨp 'ikhshap — (Sp) Yerba de la Víbora

(I) shiyamsh 'ikhshap

(V) ch'atɨshwɨ 'ikhshap

The Chumash names provided by Harrington's consultants were translated from the Spanish, meaning "medicine of the rattlesnake." These are more like descriptive terms than true names. In Mexico, for example, a dozen plant species in eight different plant families are known as *yerba de la víbora* (Martínez 1979:438–439).

In the early days, Spanish Californians and Indian people valued this plant highly as a remedy for rattlesnake bite and usually applied it in the form of a poultice (Bingham 1890:35; Blochman 1893b:232; Bard 1894:9; Jepson 1925:703; Benefield 1951:21). In 1861, Alexander Taylor wrote that the Indians of Saticoy and several parts of Santa Barbara County used *yerba del viboro* [*sic*] in snake charming. They would capture a rattlesnake and secure it in a safe place. For five or six days they fasted, drinking and bathing in a strong decoction of the plant and also chewing it. Then the charmer would teach the snake to dance, come to him, and wind itself all around his neck, arms, and body; it would even bite him without causing harm (Heizer 1973:77).

These reported uses of *Daucus* for snake control resemble what Harrington's consultants described for *chuchupate (Lomatium californicum)*, which they said was different from *yerba de la víbora*. They did not mention any connection between *yerba de la víbora* and rattlesnakes, but did give medicinal uses for the plant. It was made into a tea to be drunk for sore throat, and the dried leaves were smoked for paralysis caused by water on the brain.

Among the Ohlone also, *Daucus pusillus* was known as *yerba de la víbora* and used medicinally. A decoction was taken internally for snakebite, itching or fever, to clean the blood, or in the early stages of a cold (Bocek 1984:251).

Liliaceae | Lily Family

Dichelostemma capitatum ALPH. WOOD

BLUE DICKS, BRODIAEA

(B, I) shikh'ő'n

(Sp) Cacomite

(V) shi'q'o

The Spanish name *cacomite* is taken from the Nahuatl word *cacomitl*, a lily-like plant with a starchy tuber that was cooked in water and eaten by the Aztecs of Mexico. Spanish-speaking people applied this name to more than one kind of plant with edible bulbs. For Chumash people, *cacomite* most often refers to blue dicks. A different plant, mariposa lily, they sometimes called "another kind of *cacomite*" (see *Calochortus*). Blue dicks, formerly *D. pulchellum*, has at times been included in the genus *Brodiaea* and is often known by the common name brodiaea. Confusion also arises from the English common name "wild onion," which several early writers used to refer to blue dicks rather than to true wild onion, *Allium* spp.

Harrington's Chumash consultants described *shiq'o'n* as having blue flowers and a root like a garlic. In June, after the flower stalks had died back, people dug the bulbs by hand or with a weighted digging stick and collected them in woven sacks. They removed the leaves and stems right where they dug the plants and took only the bulbs back to the village. They usually roasted the bulbs in the hot ashes of a cooking fire inside the house.

On the islands, several Chumash families would harvest large quantities of brodiaea bulbs and roast them together in a specially constructed oven about four feet long and three feet wide. They would build a fire in this pit, and after the coals had burned down, spread the bulbs out over the coals in a layer several inches thick. They placed more hot coals and ashes on top and covered the entire oven with earth, leaving no holes for air to enter. After the bulbs were roasted, the oven was opened and a delegated person scooped them out with an abalone shell dish, putting them into receptacles for each of the families who had contributed to the harvest. Fernando Librado had seen or heard about at least four of these communal brodiaea roasting pits on Santa Cruz Island, some as large as six feet square. These bulbs may have been a particularly important terrestrial food resource for the island Chumash populations.

Shiq'o'n appears in many Chumash stories. In one moralistic tale, Coyote came to Dixie Thompson Springs in Santa Barbara, sat down, and watched children digging *cacomites*. Some children gave him some

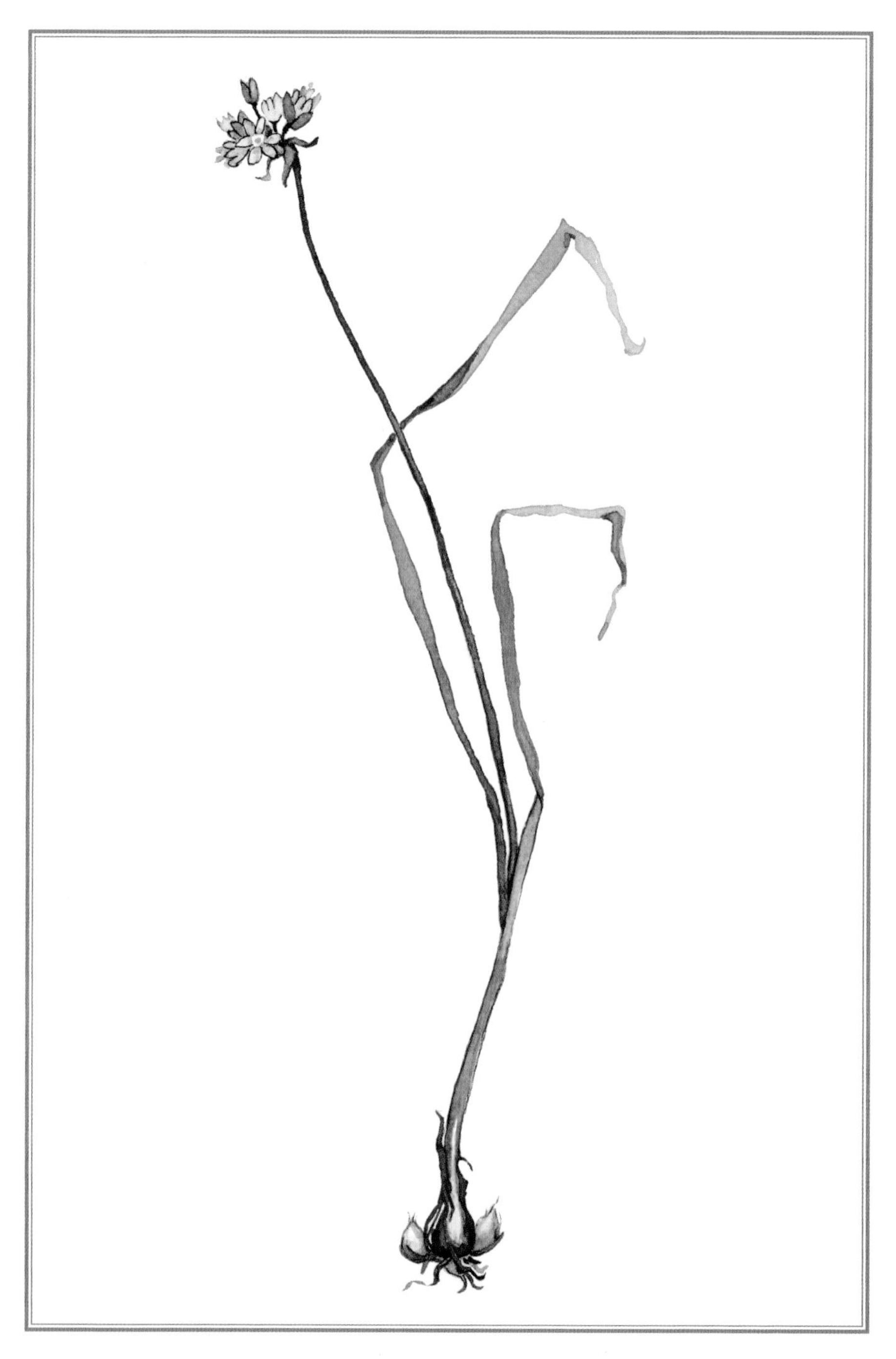

BRODIAEA
Dichelostemma capitatum

to eat; others did not. He ate raw *cacomites* until he was satisfied. Then he bewitched the bulbs in the ground so that the generous children hardly had to dig at all, but the stingy children had to dig very deeply for theirs. When the children had obtained all the bulbs they wanted, they went up to Mission Canyon to roast them. Coyote went along, too. While the bulbs were roasting, Coyote bewitched them again. When the oven was opened, the *cacomites* belonging to the stingy children were burned, but those of the generous children were perfectly cooked. Coyote and the generous children ate their bulbs while the others went hungry (Blackburn 1975:221).

Poaceae — Grass Family

Distichlis spicata (L.) E. GREENE — **SALT GRASS**

(I) lit'on — (no Spanish name known)

(V) saha

Many Indian groups utilized salt grass as a source of salt. Though *Distichlis* is common in salt marshes near the coast and inland, it seems not to have been much used by the Chumash. Perhaps they obtained salt from marine sources instead, or simply ate little salt in their food. The word for salt was *tip* (B, I). The Chumash noted whites' proclivity for salt consumption.

María Solares of Santa Ynez observed the Tulareño (Yokuts) process of gathering salt. Little girls took sticks out into the salt grass patch and switched them back and forth through the plants. The salt stuck to the sticks, and the girls scraped it off with their hands. When they had gathered a quantity of salt, they molded it into a ball that looked and tasted like pepper.

Salt grass was particularly important to the Yokuts (Latta 1977:436–442; Hoover 1973b:27–28). According to Latta, the salt grass plants were cut or pulled up and dried, then beaten with poles to remove the crystals from the leaves. The Tubatulabal followed this method and also obtained salt by sweeping the plants with willow switches early on summer mornings (Voegelin 1938:19), much as María Solares described for the Tulareño. The Cahuilla sometimes collected salt by beating *Distichlis*, but more often burned the plant and used the ashes as a condiment (Bean and Saubel 1972:66). Salt was an important trade item among many California groups.

Primulaceae | Primrose Family

Dodecatheon clevelandii E. GREENE | **SHOOTING STAR**

(B) stɨq' 'iwaq'aq' | (Sp) Ojo de Rana

Lucrecia García labeled several shooting star specimens with descriptive Barbareño Chumash and Spanish names that mean "frog's eye." She did not mention any uses for this plant. Some California peoples cooked and ate shooting stars' roots and leaves, used the flowers for women's dance ornaments, or hung the flowers on cradles to make their babies sleepy (Mead 2003:151).

Polypodiaceae | Fern Family

Dryopteris arguta (KAULF.) MAXON | **COASTAL WOOD FERN**

Pentagramma triangularis (KAULF.) YATSK., WINDHAM & E. WOLLENW. | **GOLDBACK FERN**

Polypodium californicum KAULF. | **CALIFORNIA POLYPODY**

(B, V) kepeye | (Sp) Yerba del Golpe

(Cr) kepey

(I) peye

As its Spanish name indicates, *yerba del golpe* was used medicinally to treat wounds, sprains, and bruises resulting from blows or falls. The Chumash and other people in the area made a tea by boiling the plant, particularly the roots, both to drink and as an external wash for the injuries (Bard 1894:7; Birabent n.d.; Gardner 1965:298; Weyrauch 1982:21). Spanish settlers also used this remedy (Blochman 1893b:232).

Dryopteris was the principal plant known as *yerba del golpe*, but Harrington's consultants gave the same Spanish or Chumash names to specimens of two other ferns (*Pentagramma triangularis*, *Polypodium californicum*). It is unclear whether these other species were used in the same way. Some botanists today place each of these three ferns into different plant families (Hickman 1993).

The Kitanemuk boiled the root of *yerba del golpe*, a fern, to drink for bruises (Harrington 1986). The Ohlone ate the rhizomes of *yerba del golpe* (*Dryopteris arguta*), and steeped the fronds in water to make a hair wash (Bocek 1984:247). Elsewhere, plants in the Evening-Primrose family are known as *yerba del golpe* (Ford 1975:337).

Elymus condensatus: See *Leymus condensatus*

Ephedraceae — Ephedra Family

Ephedra californica S. WATSON — **DESERT TEA**

E. viridis COV. — **MOUNTAIN TEA**

(B) woshk'o'loy — (Sp) Cañutillo

(V) kɨwɨkɨw

The same Chumash and Spanish common names refer to both *Ephedra* and horsetail (*Equisetum*), which are similar in appearance with jointed, leafless stems. (The ancient Roman natural philosopher Pliny also lumped both plants under the Greek name *ephedra*.) Harrington's consultants clearly realized the plants were different, but it is often difficult to determine from his notes which species they were talking about. Based on descriptions provided in the notes, I have attempted to separate the two genera insofar as possible.

Lucrecia García identified a specimen of *Ephedra* as "the medicinal kind" of *cañutillo* but said a sample of *Equisetum* had "no use." *Ephedra* grows primarily in the arid inland mountain areas, and the difficulty of obtaining it might have enhanced its reputed value.

According to Harrington's Chumash consultants, a decoction of *Ephedra* was good for washing cuts. They also took it internally to purify the blood and to treat unspecified "hidden illnesses" of both men and women. Other sources report that the decoction of *cañutillo* (usually identified as *Ephedra*) stems or roots was drunk cold whenever desired for kidney and bladder disorders and applied to stop bleeding from wounds (Bard 1930:22; Birabent n.d.; Gardner 1965:299). Spanish Californians later adopted the use of *Ephedra* as a blood purifier and tonic (Blochman 1894c:162). Bundles of *Ephedra* stems have been found in cave caches in the Chumash backcountry (Grant 1964:16), possibly indicating that this plant was valued medicinally from prehistoric times.

Ephedra has continued to be used in similar ways. The green branches, collected in spring, are dried and stored indefinitely, then steeped in water to make a wine-colored tea taken to purify the blood and urinary system (Weyrauch 1982:9).

Chumash consultant Fernando Librado described "painting" three decorative rings around an arrow shaft, just below the feathers, with *cañutillo*. The dried stem was wrapped once around the arrow and pulled tight. This left a mark, the color of which was "according to the color of the plant." This *cañutillo* with which arrows were marked was probably *Ephedra*, since the Navajo use a species of *Ephedra* as a dye for wool (Niethammer 1974:97) and basketry material (Elmore 1943:24). It seems unlikely that *Equisetum* could have served this purpose since it is not particularly rich in pigment, although a gray dye can be made from it (Krochmal and Krochmal 1974:249).

In New Mexico, both *Ephedra* and *Equisetum* were known as *cañutillo* and used to treat venereal disease (Ford 1975:155). Throughout the arid Southwest, however, *Ephedra* was by far the more important medicinal plant, taken mainly as a tea. It was used to treat venereal disease, fevers, kidney and urinary complaints, and stomach disorders (Ford 1975:155). The medicinal use of *Ephedra* was also common among other California Indians (Strike 1994:56). The Cahuilla took a tea of boiled *Ephedra* twigs as a blood purifier, for stomach and kidney ailments, and for venereal disease (Bean and Saubel 1972:70–71).

Onagraceae — Evening-Primrose Family

Epilobium canum (E. GREENE) RAVEN — **CALIFORNIA FUCHSIA**

(B) s'akht'utun 'iyukhnuts — (Sp) Balsamillo, Balsaméa

This plant (formerly *Zauschneria californica*) was used medicinally for cuts, sores, and sprains, particularly in livestock. The dried, powdered leaves were sprinkled directly on the wound, or boiled as a wash or bath. Though it was mentioned by several authors who worked in this region (Birabent n.d.; Benefield 1951:24; Blochman 1893b:231–232; Saunders 1933:142), this remedy may not be indigenous among the Chumash.

The Chumash term means "hummingbird sucks it," describing the tubular red flowers. Other plants with similar flowers were also called by this name.

Equisetaceae — Horsetail Family

Equisetum spp. — **HORSETAIL, SCOURING RUSH**

(B) woshk'o'loy — (Sp) Cañutillo

(V) kɨwɨkɨw

Because the rough, corrugated stems of horsetail have high silica content, they can be used as sandpaper. The Chumash polished their wooden bowls with dry horsetail stems, then painted them with red

CALIFORNIA FUCHSIA
Epilobium canum

ochre and set them in the sun to let the oil binder in the paint soak in. Many other California Indian groups used *Equisetum* like sandpaper for polishing arrows and woodwork (Mead 2003:158–161). (A Chumash method of marking hunting arrows with *cañutillo* is described under *Ephedra*.)

As noted above, *Equisetum* and *Ephedra* somewhat resemble one another in having green, tubular, jointed stems without apparent leaves. Since the two genera also share Chumash and Spanish common names, there is some difficulty in distinguishing them in Harrington's notes. Both had medicinal uses and, although *Ephedra* was preferred, some Chumash seemed to regard *Equisetum* as an acceptable medicinal substitute in certain cases. Horsetail did have the advantage of being more commonly available near the coast.

Although one consultant said that horsetail was of no use, another said that the tea could be drunk as a purgative when nothing better was available. *Equisetum* was considered suitable for treating "hidden illnesses" of men and women. Some of the cañutillo remedies for bladder and kidney disorders listed under *Ephedra* may have employed *Equisetum.*

Of the two *cañutillos, Equisetum* is less widely used in medicine. Both plants were used by New Mexico Spanish people to treat venereal diseases (Ford 1975:155). The Cahuilla have a story about a culture hero who ate some *Equisetum* stalks, vomited them, and became well (Bean and Saubel 1972:70–71), suggesting that the Cahuilla may have used horsetail as an emetic.

Asteraceae — Sunflower Family

Ericameria arborescens (A. GRAY) E. GREENE — **GOLDEN FLEECE**

(no Chumash name known) — (Sp) Yerba del Pasmo

The plant most widely known as *yerba del pasmo (Adenostoma sparsifolium)*, a member of the rose family, has a relatively restricted distribution in Chumash territory. Some investigators have suggested that *A. sparsifolium* was overcollected for medicinal use and virtually wiped out by Hispanic settlers in the region (Barber 1931). Chumash consultants identified specimens of *Ericameria* (formerly *Haplopappus arborescens*) as "*yerba del pasmo*," which may not be a case of mistaken identity as some have assumed (Smith 1998:158), but an indication that this more common species was considered an acceptable form of an increasingly rare medicinal plant (Timbrook 1987:178). Kitanemuk people today gather *Ericameria arborescens* for medicinal use, identifying it as *yerba del pasmo* (Caterino and Juanita Montes, personal communication 2006). See *Adenostoma sparsifolium.*

Hydrophyllaceae — Waterleaf Family

Eriodictyon crassifolium BENTH. — **YERBA SANTA**

(B) wishap' — (Sp) Yerba Santa

(I, V) wishap

The Spanish missionaries gave various members of the genus *Eriodictyon* the name *yerba santa*, or "holy herb," in recognition of the plants' medicinal value. Specimens collected by Harrington's consultants indicate that although several species of *Eriodictyon* occur in this area, only *E. crassifolium* was called *yerba santa* by the Chumash. They boiled its leaves and drank the tea for colds, chest pain, cough, and fever. They used it externally as well, to soothe foot pain, chest pain, and other ailments, applying hot baths of *yerba santa* and rubbing the affected parts.

In the late nineteenth century, settlers as well as Indians made medicinal use of *yerba santa* (Rothrock 1878:46–47; Bingham 1890:36; Bard 1894:5; Birabent n.d.; Harrington 1986; Benefield 1951:21, 25). They simmered the leaves in water and drank the resulting liquid warm for ailments of the throat and lungs, including coughs, colds, and asthma. They also infused the leaves in alcohol or whiskey to be taken internally or applied as a liniment, a method mentioned by Harrington's consultants.

Chumash descendants and others have continued to take *yerba santa* for respiratory ailments, to expel mucus, and as a general tonic. As a remedy the tea must be drunk at frequent intervals, and many people add sugar to counteract the very bitter taste (Weyrauch 1982:22).

Traditionally, *yerba santa* seems to have conferred some degree of supernatural power. A Chumash healer's medical kit contained two stones that were believed to be "alive" because they had been treated with oil from an eel mixed with *yerba santa* (Walker and Hudson 1993:50). A person could ensure victory in a game of shinny by making a special ball from *yerba santa* root pounded and mixed with the blood of some poisonous animal such as a Jerusalem cricket (Hudson and Blackburn 1986:393).

Yerba santa appears in Chumash place names recorded by Harrington in the upper Santa Ynez Valley, near the mouth of the Ventura River (Applegate 1975b:33, 45), and near San Roque in the Santa Barbara foothills. These might denote spots where the plant occurred.

Several *Eriodictyon* species—called *yerba santa* or *palo santo*—have been widely used to treat numerous health complaints, particularly rheumatism, respiratory problems, and wounds, and to purify the blood. The Ohlone (Bocek 1984:235), Salinan (Garriga 1978), Cahuilla (Bean and Saubel 1972:71), and Paipai (Ford 1975:352–353) were among the many indigenous groups who used this plant.

Polygonaceae — Buckwheat Family

Eriogonum elongatum BENTH.
Eriogonum nudum BENTH.
(V) 'an

WILD BUCKWHEAT

(Sp) Tibinagua

Tibinagua—a folk category that may include several *Eriogonum* species—was employed medicinally by the Chumash and others in the region. The Chumash used a tea of wild buckwheat, *'an*, to stop hemorrhage (Blackburn 1963:147). To purify the blood, they made a decoction of *tibinagua* leaves to drink cold during the day, whenever they felt thirsty (Birabent n.d.). Santa Ynez Chumash made a tea of *tibinagua* roots (identified as *E. elongatum*), taken cold or hot with sugar, as an effective remedy for fever (Gardner 1965:298).

The Ohlone referred to *Eriogonum latifolium* as *tibinagua*. They drank a decoction of *tibinagua* roots, stems, and leaves for colds and coughs (Bocek 1984:249). The Kawaiisu also drank a hot infusion of *tibinagua* roots (*E. nudum* var. *pauciflorum*) for colds and coughs (Zigmond 1981:30).

Polygonaceae — Buckwheat Family

Eriogonum fasciculatum BENTH.
(V) tswana'atl 'ishup

CALIFORNIA BUCKWHEAT

(Sp) Poléo

Harrington's Chumash consultant Lucrecia García confidently identified specimens of California buckwheat as *poléo*. This is unusual: elsewhere the name *poléo* has generally referred to certain strongly aromatic members of the mint family, often mint (*Mentha*) or pennyroyal (*Monardella* spp.). California buckwheat superficially resembles *Monardella* when flowering but has no pronounced scent. Although confusion between the two plants seems unlikely, some of *poléo*'s medicinal uses reported here may actually pertain to *Monardella*.

For rheumatism and irregular menstruation, the Chumash boiled the whole top of the *poléo* plant, drank the tea, and also bathed with it. *Poléo* combined with sage was reportedly used to suppress menstruation (Bard 1894:6). For stomach trouble the plant was simmered and the decoction drunk hot (Birabent n.d.); the flower could also be used in this way (Gardner 1965:298).

In Harrington's Salinan collection, a specimen labeled "*poleyo*" was a species of *Monardella*. The Ohlone made decoctions and salves of *poléo (Monardella villosa)* for respiratory problems (Bocek 1984:253). Among most southwestern peoples, *poléo* referred to any of several species of mint (*Mentha*) or false pennyroyal (*Hedeoma*) and was used in the treatment of fever and menstrual and stomach problems (Ford 1975:277–278).

Geraniaceae — Geranium Family

Erodium cicutarium (L.) L'HÉR. (Introduced) — **REDSTEM FILAREE**

E. moschatum (L.) L'HÉR. (Introduced) — **WHITESTEM FILAREE**

(B) chikwɨ', s'u'wlima' — (Sp) Alfilerillo

(I) s'u'wlima'

(V) kwɨ'ɨn

Both these species are widespread weeds that were introduced to California early in the mission period; *E. cicutarium* was particularly esteemed as livestock fodder (Jepson 1936:406). Chumash consultants reported using the seeds of *alfilerillo* for food, and they boiled the whole plant to make a tonic for the blood. Modern Chumash people apply the name "*chikwit*" (Barbareño *chikwɨ'*) to a different introduced weed, scarlet pimpernel (*Anagallis arvensis*), which they also use medicinally (Gardner 1965:298; Weyrauch 1982:8).

Elsewhere in California, the Yokuts, Huchnom, and Nisenan ate *E. moschatum* as greens (Mead 2003:172–173). In the southwestern United States and northern Mexico, *alfilerillo (E. cicutarium)* is used as a diuretic and for urinary disorders, venereal disease, body sores, sore throat, and rheumatism (Ford 1975:125–126). The Ohlone ate the raw stems and treated typhoid fever with a tea of the leaves (Bocek 1984:252).

Papaveraceae — Poppy Family

Eschscholzia californica CHAM. — **CALIFORNIA POPPY**

(B, I, V) qupe — (Sp) Toroza

California poppy figures in Chumash beliefs and myths. According to Fernando Librado, people used to say that poppies were the ruin of girls. Boys would take girls out gathering poppies, and the flowers' beauty would overcome the girls and cause them to yield to the boys (Blackburn 1975:171). One story, from the time when animals were people, recounts Lizard describing to Coyote the bright yellow and orange poppies out on the islands: "When you see it, it is as if the sun itself is on the ground, so beautiful is the flower" (Blackburn 1975: 158).

After death, as the soul journeys to *Shimilaqsha*, the Land of the Dead, it must pass a place where two ravens peck out its eyes. There are many poppies growing there in the canyons on each side. The soul quickly reaches out its arms to each side, picks two of these poppy flowers, and puts one into each eye socket. Thus it is able to see again immediately. When the soul finally gets to *Shimilaqsha*, it is given eyes of blue abalone (Blackburn 1975:100).

None of Harrington's consultants mentioned any medicinal uses for California poppy. Bard, however, claimed that it had analgesic properties and was used for colic (Bard 1894:6).

Among the Ohlone, a decoction of poppy flowers was rubbed into the hair to kill lice. One or two of the flowers were placed under the bed to put a child to sleep. Pregnant or lactating women avoided the plant, believing its scent to be poisonous (Bocek 1984:251). Some other California peoples ate the leaves and used the roots medicinally (Heizer and Elsasser 1980:245; Strike 1994:61), such as for toothache, headache, and stomachache (Chesnut 1902:351).

Oleaceae — Olive Family

Fraxinus dipetala HOOK. & ARN. — **FOOTHILL ASH**

(B, I) wɨntɨ'y — (Sp) Fresno

(V) wɨltɨ'y

The two principal Chumash uses of foothill ash were as pigment and as a beverage. People burned the wood of this small tree to produce a very fine charcoal, which they ground and mixed with water to make black paint. This charcoal may be one of the pigments used in Chumash rock art.

CALIFORNIA POPPY
Eschscholzia californica

They also made a refreshing drink by steeping the twigs of foothill ash until the water turned bluish in color. This was particularly good medicine for sick people, for it was very cooling. Harrington's consultants said the Chumash first learned of its value when a dying Indian lay down under an ash tree and recovered. The priest at Mission San Fernando in 1814 mentioned this as a remedy of the local Indians (Geiger and Meighan 1976:73).

Both these uses of *Fraxinus* were reported among the Ohlone (Bocek 1984:250). *Fresno* was also made into a refreshing drink by the Salinan (Garriga 1978:28) and in Mexico (Ford 1975:198). The Pomo poulticed wounds with *Fraxinus* roots (Mead 2003:182).

Asteraceae — Sunflower Family

Gnaphalium bicolor BIOLETTI — **BICOLORED CUDWEED**

G. californicum DC. — **GREEN EVERLASTING**

G. canescens DC. — **WHITE EVERLASTING**

(no Chumash name known) — (Sp) Gordolobo

These three kinds of everlasting were called *gordolobo* and used medicinally by the Chumash, but Harrington's consultants did not provide further details. Bundles of the plant have been found in dry cave caches in the Cuyama and Hurricane Deck regions of Santa Barbara County (Grant 1964:16). *Gordolobo* was reported to have been used in this region for pulmonary diseases (Bard 1894:22) or to expel gas from the stomach (Birabent n.d.).

"*Gordolobo*" has also been reported as a Californian Spanish name for an unrelated plant, common mullein (*Verbascum thapsus* L.), which was used for sprains and pulmonary diseases (Ford 1975:203). The *gordolobo* discussed by Bard (1894:22) may be *Verbascum* rather than *Gnaphalium*.

The Ohlone, Salinan, and Miwok used *Gnaphalium californicum* for colds and stomach pains (Bocek 1984:254; Garriga 1978:25; Mead 2003:190). Other uses reported for the various *Gnaphalium* species in California and northern Mexico include eyewash, poultice for swelling, and tea for heart pains (Strike 1994:67; Ford 1975:203).

Asteraceae — Sunflower Family

Grindelia camporum E. GREENE — **GUM PLANT**

(V) stɨq shi'sha'w — (Sp) Copaiba

According to one Harrington consultant, *copaiba* was the name of a plant that grew on the islands and had big, yellowish flowers which were very gummy. It was made into a medicinal "balsam" or salve and sold in drugstores. He did not specify the purpose for which this medicine was intended.

Grindelia robusta (now *G. camporum* E. Greene) was reportedly used in the Chumash region to treat poison-oak rash, skin diseases, and pulmonary troubles (Bingham 1890:34; Bard 1894:6). The fresh plant was applied to the affected area or used in the form of a decoction or alcohol infusion. The early botanical explorer Rothrock also noted that Californians used either the gummy exudate or a tincture made from this plant to treat poison-oak, though he was skeptical of its efficacy (1878:46).

Other species of *Grindelia* were used for poison-oak dermatitis, boils, and wounds by the Ohlone (Bocek 1984:254); to cure colds by the Cahuilla (Bean and Saubel 1972:75); and by Hispanic people in the Southwest for colds, rheumatism, kidney disorders, paralysis, and stomach trouble (Ford 1975:269).

Asteraceae — Sunflower Family

Helenium puberulum DC. — **SNEEZEWEED**

(V) manakhshmu — (Sp) Rosilla

Harrington noted that the Chumash used *rosilla* medicinally for influenza. According to other sources, the typical method was to powder the flowers and take them as a snuff to treat head colds, catarrh, or influenza (Bingham 1890:35–36; Birabent n.d.; Gardner 1965:298). The plant was also taken as a tonic and as a treatment for scurvy (Bingham 1890:35–36; Blochman 1894c:163).

Fernando Librado told Harrington that a famous pipe doctor, Iluminado, employed powder of the *rosilla* flower to cure the effects of tobacco smoking. When treating a sick person, Iluminado would take three or four draws on the pipe without letting out any smoke, and then blow it out all at once right into the patient's face. Afterwards, to cure himself from the evil effects of such smoking stunts, Iluminado crushed the dry *rosilla* flower in his hands, mixed it with cold water, and drank it.

The Ohlone also used this plant to treat colds and sprinkled the powder on wounds (Bocek 1984:254). Other species of *Helenium* were used for many medicinal purposes in the greater Southwest (Ford 1975:331, 339–340).

Asteraceae — Sunflower Family

Hemizonia fasciculata (DC.) TORREY & A. GRAY — **TARWEED**

(B, I, P, V) swey — (Sp) Escoba Amarilla

Tarweed is one of several plants used to make brooms. All such plants might be called *escoba*, or "broom" in Spanish, although they were differentiated by separate terms in Chumash languages. Several plant specimens of *Hemizonia fasciculata* were labeled "*swey*, *escoba amarilla*," but these vernacular names may have included more than one botanical species. Another tarweed, *Hemizonia increscens*, becomes far more common north of Gaviota. See also *Lotus scoparius*, *Rhus trilobata*, and *Symphoricarpos* for other plants called "*escoba*."

Fernando Librado said that in the old days they gathered *escoba amarilla* plants two to three feet long and tied them in bunches to use for sweeping. They considered the larger plants to be too brittle for this purpose. In historic times, they tied the tarweed bunches to a stick handle to make brooms for sweeping outside the missions or for cleaning the bake ovens.

Hemizonia produces small black seeds that the Chumash gathered for food. They dried the seeds, rubbed them between the hands, and winnowed away the chaff. The island Chumash winnowed with an abalone shell rather than a basket; they shook the shell and as the chaff rose to the surface they blew it away. After winnowing, they pounded the seeds in a mortar and mixed in a little water to form a ball, which was usually eaten raw. Some people toasted the tarweed seeds with hot coals in a basket before pounding them, but toasting made the seeds very oily. Seeds of other tarweeds, such as *Madia* spp., may also have been eaten in this way.

In early times this tarweed seed meal was commonly eaten, but it had virtually gone out of use by Harrington's day. When he interviewed Luisa Ygnacio in Santa Barbara about 1913, she remembered having seen it prepared only once, when José Venadero brought some for her mother-in-law, María Ygnacio, to fix. Luisa said it was like a black, dry flour of "not disagreeable" taste.

Chumash place names that refer to this plant are still in use today. *'Aliswey*, a village "in the tarweed" (I) northeast of Los Alamos, is seen today as Lisque Creek. Both Suey Crossing and Rancho Suey were named for *Swey*, "tarweed" (P), a village east of Santa Maria (Applegate 1975b:25, 43).

Several California peoples made *Hemizonia* seeds into pinole, a pulverized meal or paste eaten either dry or with water added to make a gruel (Mead 1972:102; Heizer and Elsasser 1980:246; Bocek 1984: 254). The Cahuilla boiled the whole plants down into a tarry gum, which they ate in times of famine (Bean and Saubel 1972:77).

Rosaceae — Rose Family

Heteromeles arbutifolia (LINDLEY) ROEMER — **TOYON, CHRISTMAS BERRY**

(B, Cr, I) qwe' — (Sp) Toyon

(O) ch'okoko, chmishɨ

(V) qwe

Toyon fruit was a traditional food that the Chumash continued to use well into historic times. The Indians at the Santa Barbara Mission used to go up into the canyon above the mission and collect toyon fruit, which they brought back in cloth sacks. The net bags used in gathering acorns and *islay (Prunus)* were too coarse for these small berries.

The Chumash said that raw toyon fruit tends to stick in the throat. They sampled it for sweetness as they picked and often did the first step in preparing it right where they gathered it. Most commonly, they toasted the fresh toyon berries in a steatite *olla* over hot coals, stirring frequently. This did not take very long, just until the berries were good and hot. According to Candelaria Valenzuela, these "hollyberries" were roasted until they turned white (Blackburn 1963:144). Some people just spread the berries in the hot sun until they turned black and then mashed them. After either of these preliminary treatments, the fruit was placed in an old water basket or some other vessel and allowed to stand "until it was good and ready." Some of Harrington's consultants said this took about three days, others five or ten. After that, the toyon fruit was soft and sweet and could be eaten without any further preparation "as one would eat beans or bread," sometimes along with toasted, ground chia. The Chumash never boiled toyon.

TOYON
Heteromeles arbutifolia

Toyon has hard wood, which the Chumash found useful for a variety of tools and utensils. Among these were fishhooks (Henshaw 1955:152), harpoons, and bone-pointed fish spears for taking salmon (Hudson and Blackburn 1982:194). Basketry awls, matting and thatching needles, digging sticks with fire-hardened points, reamers, wedges, and hide scrapers were made of toyon. The Chumash also used this wood to make wooden pestles, bowls, drinking cups, walking sticks, war clubs, canoe pegs, and cradle frames. They mounted drill bits and chisels into hafts or handles of toyon. Instead of the usual arrow shaft straighteners of stone, a specialized, grooved tool of toyon wood was employed to straighten reamers, elderberry sticks for flutes, and any shafts longer than arrows. The piece to be straightened was softened with heat or water, placed into the perfectly straight groove worked in a long stick of toyon, and lashed in place until it dried (Hudson and Blackburn 1987:106). The Chumash made various gaming implements of toyon wood, including shinny sticks, balls, and poles for hoop-and-pole games. They smoked fish on toyon-stick racks and also burned the wood as fuel for smoking dried fish.

Harrington's consultants provided detailed information about arrow-making (Hudson and Blackburn 1982:96–124). Toyon shoots were made into arrows, particularly the sharpened one-piece arrows used in hunting or competitive shooting. The arrow maker selected shoots one-quarter inch to one-half inch in diameter and cut them all to the same length, about two to three feet. In order to straighten them, he steamed bundles of these rods between layers of wet grass buried in hot coals. It was important to straighten each of these wooden shafts before it dried; some required a second steam treatment. After straightening a shaft he cut off all the knots, then sharpened the point and hardened it in hot ashes while the wood was still green. Finally he attached three split feathers to the nock end with sinew and tar. The hardwood foreshafts of compound cane arrows, with or without a stone point attached, could also be made using toyon wood. Cane arrow shafts, unlike the wooden ones, were dried before straightening. See *Arundo*, *Leymus*, and *Phragmites*.

Some consultants mentioned other weapons made of toyon wood. Although Fernando Librado considered the wood too heavy for this purpose, some Chumash did make both sinew-backed and unbacked (self) bows from toyon. Using three-foot-long toyon sticks to propel rocks, boys could hit a target at forty paces (Hudson and Blackburn

1982:141). A type of rabbit stick made from toyon may have been either thrown or held as a club to kill small game.

In Chumash ceremonial activities, toyon played a role in dance regalia and offertory poles. The *tsukh*, or feathered crown headdress, was held in place with a sharpened wooden pin inserted from the right side. This headdress pin was made from a toyon stick about a quarter inch in diameter and fifteen inches long, painted red. At the outer end a bunch of feathers, split lengthwise to make them more flexible, was attached with tar and cordage.

Toyon or *palo colorado (Ceanothus spinosus)* wood made the best offertory poles because they lasted well in the ground and would not rot. One such feathered offertory pole marked a shrine near Ventura called the Depository of the Things of the Dead. Fernando Librado, who saw this pole, said the bark had been peeled from a straight pole about five and a half feet long and the lower end sharpened to a point. When set into a hole lined with stones, this offertory pole was about four feet high. They had painted it with a mixture of red ochre and a kind of black rock ground up together. There were various kinds of feathers—crow, gull, and other birds—tied onto the pole in a bunch. At the top, two vulture feathers stood straight up, the insides of the feathers turned to the west and the faces to the east. During the winter solstice ceremony, dancers held two toyon sticks, similarly ornamented with feathers, throughout five days of dancing. They then took the dance sticks up to this shrine and deposited them there. People also left offerings of seeds, beads, tobacco, money, clothing, and the like there on other occasions.

The common name toyon is from a Spanish adaptation of the Ohlone Indian word for the plant, *tottcon* (Harrington 1944:37). The Ohlone ate toyon berries roasted or dried (Bocek 1984:249). In much the same way described for the Chumash, several other California Indian peoples "wilted" toyon berries with heat, boiled or roasted them, or allowed them to soften over a period of time before eating them (Mead 2003:203).

Asteraceae | Sunflower Family

Heterotheca grandiflora NUTT. | **TELEGRAPH WEED**

(no Chumash name known) | (Sp) Yerba de la Pulga

One Chumash consultant said this strong-smelling plant was used to repel fleas.

Rosaceae | Rose Family

Horkelia cuneata LINDLEY | **WEDGE-LEAVED HORKELIA**

Potentilla glandulosa LINDLEY | **STICKY CINQUEFOIL**

(B) chiqwi 'ikhakha'kh | (Sp) Tabardillo

According to Harrington's consultants, the root of the plant called *tabardillo* was boiled and the tea drunk for fever and colds. They also considered this plant to be good for the stomach. By the 1950s, a few Chumash recalled that *tabardillo* root was ground or mashed, boiled, and taken for the blood (Gardner 1965:298). Several other sources confirm the early use of the plant by both Indians and settlers to treat fever and Spanish flu, as a blood purifier, and as a stimulant (Bard 1894:6; Birabent n.d.; Benefield 1951:24).

The botanical identity of the plant known as *tabardillo* is uncertain. In the Harrington collection at the National Anthropological Archives, specimens of *Horkelia cuneata* and *Potentilla glandulosa* were labeled "*tabardillo*." Other sources, such as Bard (1894:6), identify *tabardillo* as a completely unrelated Mexican ornamental plant, *Ageratina adenophora* (formerly *Eupatorium*), that has become naturalized as a weed in the south coast area. The Chumash term means "big *alfilerillo*," referring to another non-native plant, filaree (*Erodium*), whose leaves somewhat resemble those of *Horkelia* and *Potentilla*. *Tabardillo* might not be an indigenous Chumash remedy.

Juglandaceae | Walnut Family

Juglans californica S. WATSON | **CALIFORNIA BLACK WALNUT**

(B) ktɨp | (Sp) Nogal [tree], Nuez [nut]

(V) tɨpk

The Chumash ate wild walnuts despite the hard shell and small nutmeats. It requires a great deal of effort to crack open the nuts using a hammerstone, but they taste very good and were considered superior in flavor to the introduced English walnuts. Indian people apparently

A ROLL OF THE DICE

Games of skill and chance were very popular with Chumash children and adults, both men and women. In this scene, two women play their favorite game, pɨ. *Singing to bring success in the game, a player tosses the walnut shell dice onto a basketry tray. If the right number of dice land flat-side-up, she wins a round and takes a counter stick to keep score. Stakes in this and other games can be high—perhaps money, valuables, or even political office.*

traded native walnuts from southern to central California, where the trees are often found near former village sites (Jepson 1909:365).

A gambling game called *pɨ* was played with dice made from wild walnut shells split into hemispherical halves and filled with tar. The Chumash sometimes decorated the dice with shell inlay, but the shell bits did not signify numbers as the dots on modern dice do. Instead, the score depended on the number of dice that landed with the flat side up. This game was both a pastime, especially of women, and a way of determining succession to political office (Hudson et al. 1977: 21–22). It was played with six dice thrown on a flat basketry tray as follows:

> Two players sit across from each other with the tray between them and a pile of ten counter sticks to one side. One player selects "odd," the other "even." One player then picks up the dice

> and tosses them onto the tray, noting the number which have fallen with the flat side up. The same player rolls the dice a second and a third time. At the end of three throws, the three scores are added to determine if "odd" or "even" won the game. If the player who had selected "odd" rolls an odd number total, she receives a counter stick and gets another three throws. But if the total is even, she receives no stick and the turn passes to the other player. After all the counter sticks have been taken from the pile, the player who wins each set of three rolls takes a counter from her opponent. When one player has won all the counters, the pi game is over (Hudson and Timbrook 1997:16).

The Chumash occasionally employed walnut bark in basketmaking. For example, some weavers finished off the rim of the small, coarsely twined utility baskets called *chiquihuites* by wrapping the warp ends with strips of walnut bark. One consultant suggested that walnut bark was also used for designs on coiled basketry (Craig 1967:111–112), but others denied this and no actual examples have been identified. Perhaps this meant it was used as a dye, but no mention of walnut as a dyestuff was found.

In more recent times, a Santa Ynez Chumash man mentioned that walnut leaves were good for making tea to drink for the blood (Robert Ontiveros, personal communication 1978). This practice was also reported among the Ohlone (Bocek 1984:248).

Juncaceae — Rush Family

Juncus spp. — **RUSH** (general)

(no generic Chumash name known) — (Sp) Junco

Chumash peoples recognized at least three types of *Juncus* but had no name for the group as a whole. Harrington's consultants sometimes specified a particular folk taxon of *Juncus* as being used in a particular way, and these are discussed separately according to the botanical species cluster to which they have been attributed. The Spanish term "*junco*" is an inclusive one, however, referring to all the species in the botanical genus *Juncus*. This section includes only Chumash uses of *Juncus* in general or those for which a species is not known.

Juncus had quite a number of uses, most importantly in basketmaking. Harrington's consultants were familiar with a wide variety of baskets that Chumash weavers traditionally made with *Juncus*. For example, a *pɨtsmu* (V), a twined shallow basket covered with tar, was carried along in a plank canoe for bailing. One type of large burden basket, called *wo'ni* (V), was bucket-shaped and made by coiling. Carried on the back in a net suspended from a tumpline, or forehead band, these burden baskets were essential in gathering and transporting foodstuffs. Small basketry dishes, *kuyiwash* (V), held water for drinking, for filling the water bottle, for soaking basketmaking materials, or for bathing babies. Food was served in these versatile basketry dishes, and *toloache (Datura)* was drunk from them as well (Craig 1967:98,108). After Spanish colonization, the Chumash used an implement made of *junco* to stir corn being toasted for pinole.

Other kinds of baskets are described under *Juncus acutus*, *J. balticus*, and *J. textilis*. In general, *J. acutus* was employed in the twined technique and the other two in coiled basketry, as foundation and sewing material, respectively. Chumash baskets have been studied extensively elsewhere (Dawson and Deetz 1965; Hudson and Blackburn 1982, 1983). Some archaeological specimens of *Juncus* baskets were described by Grant (1964).

The Chumash had a number of other uses for *Juncus* besides basketmaking. In the final stage of making shell-bead money, the rough edges of the beads were polished down to make them perfectly round. To do this, the bead maker split a foot-long section of *Juncus* stem, cleaned it of pith, and trimmed it. The *junco* had to be just the right size, so that when the drilled beads were strung onto it and packed together, they fit tightly and did not turn independently. The bead maker sat and rolled the rod of beads on a specially made thin slab of rock held in the lap. About three inches' worth of beads were polished on the *junco* rod at one time.

In the old days, the Chumash boiled abalones whole in salted water and then dried them for later use. They strung the abalones on a whole stem of *junco* that had been picked two or three days beforehand. They hung strings of dried abalones like clotheslines.

Using *Juncus* needles, the Chumash made mats by stringing together pierced stems of tule (*Scirpus* sp.). They took a foot-long piece of *junco* stem, cut it to a point at the lower (root) end, where it was strongest, and finely shredded the other end. It was a good idea to make many of these needles at a time, cutting the *junco* and shredding the ends while it was still green, then laying the needles aside for future use. The cordage to be used in stringing the mats was unraveled slightly, spliced onto the shredded end of the *junco*, and trimmed of projecting ends. The tule stems were pierced with a needle of wood, then the sharp end of the *junco* needle was inserted and threaded through, carrying the cordage along behind it. Mats were also made of *Juncus* stems held together by twining; it was not necessary to pierce them.

Some items of apparel were made of *Juncus*. An apron of *junco* or of *carrizo (Leymus, Phragmites)*, extending down nearly to the knees, was worn under the belt made for carrying game. For the Fox Dancer's headdress, the Chumash twined *Juncus* to form a cap, bringing the ends of the *junco* together behind the dancer's head and working them into a long, tail-like braid which was weighted with a stone at the end (see Hudson et al. 1977:69–70). Sometimes they wove a fox head of *junco*, covered it with grass or leather, and painted it to resemble the Fox.

Throughout most of the central and southern California mission area, *Juncus* was a common material for both twined and coiled basketry (Harrington 1942:23), but few other peoples used it to the extent the Chumash and the neighboring Gabrielino and Fernandeño did. These three groups were the only ones to make coiled baskets with a three-rod *Juncus* foundation (Dawson and Deetz 1965:207).

Various California Indian groups used *Juncus* spp. in other ways as well. The Southwestern Pomo held shell beads on *Juncus* stems while rolling them on a stone slab; the Luiseño and Cahuilla made *Juncus* mats for storing ceremonial articles; and the Maidu wore breechcloths of *Juncus* (Mead 2003:213–214; Bean and Saubel 1972:80–81).

Juncaceae — Rush Family

Juncus acutus L. — **SPINY RUSH**

J. effusus L. — **BOG RUSH**

(B, I) 'ulat' — (Sp) Junco

(V) 'esmu

The Chumash classification of some plants did not correspond exactly to the categories we recognize. Today's taxonomic system is based primarily on plants' reproductive structures, their flowers and fruit. In the case of *Juncus*—"rush" in English, or "*junco*" in Spanish—the Chumash recognized species based on overall growth habit and features of the stems. Chumash folk taxa may therefore include elements of more than one botanical species.

The plant known as *'esmu* (V) or *'ulat'* (B) was the kind of *Juncus* used in twined basketry. It included *Juncus acutus*, possibly *J. effusus*, and perhaps some individuals of other species (*J. patens?*) as well. Harrington's consultants described this type as having strong but slender stems with sharp, needlelike points, and a habit of growing in clumps or bunches. It grew abundantly at Mugu between the estuary and the cliffs, at Hueneme, and at Ventura, they said, but not around Santa Barbara.

According to Fernando Librado, *'esmu* stems were not cut but pulled from the plant, one by one, and used immediately after picking. The stems were used whole, not split, in making several types of twined baskets.

Water bottle baskets—called *'awa'q* (B, I) or *'ush'e'm* (V)—were of various sizes; some surviving examples have a capacity of a gallon or more. Chumash peoples usually caulked their water baskets with asphaltum. First, they put a thick coat of mud over the outside of the finished basket. Then they pulverized hard mineral tar and put it into the bottle, followed by several small stones which had been heated in the fire. They rotated the basket to roll the stones around inside it, melting the tar and coating the inside of the basket, thereby making it watertight. Any tar that oozed out through the weave during this process bonded with the mud on the outside, making an even more impervious container. Some people preferred to wash off the mud afterwards, but that was not necessary. The basket was filled with water, allowed to stand overnight, and then washed out, after which it was ready for use.

For gathering clams and things to be used as bait, the Chumash made twined openwork baskets of *'esmu*. These carrying baskets were about a foot across and a foot high and had a handle at the top. Being open-meshed, they acted as strainers and were also useful for carrying fish. They were generally known as *'okhoy* (V) or *wachka'y* (I), but larger ones, with a reinforcing rod at the rim, were sometimes called *tsaya* (V, I).

The term *tsaya* (V, I), or *chaya* (B), usually referred to a basketry tray used for leaching acorn meal. This type of basket was also twined from whole *'esmu* stems. Some consultants said these baskets were closely twined so the meal would not fall through, but others said they were openwork with the meshes half an inch wide. Perhaps in the latter case a lining of leaves was required, though none of the consultants mentioned it. The leaching basin was about two feet in diameter and reinforced by lashing a willow rod around the rim. To leach acorn meal, one would place the basket on rocks so that it would be held stable about six inches above the ground, and then spread the acorn flour over it in a layer about an inch and a half thick. Rosario Cooper preferred to sprinkle the meal around the edge of the basin in a sort of spiral form. She slowly poured water over this, and as it trickled through, it washed the bitter tannic acid from the meal. Not everyone had these special leaching baskets; some people leached acorn flour in other kinds of vessels or in hollowed depressions in sandy ground.

These types of Chumash twined baskets are described in Dawson and Deetz (1965), Craig (1967), and Hudson and Blackburn (1982, 1983). Contrary to Merrill (1923:230, 237), *Juncus acutus* was not the only plant species the Chumash used in twined basketry; see other species of *Juncus* listed below, and also *Rhus trilobata*, *Salix*, and *Scirpus*.

Juncaceae — Rush Family

Juncus balticus WILLD. — **WIRE RUSH**

J. textilis BUCHENAU (smaller stems only) — **INDIAN RUSH**

(V) tash — (Sp) Junco

As I have noted, Chumash folk species often differ somewhat from those usually recognized by botanists today. It is not clear whether an equivalent category existed in Barbareño or Ineseño Chumash, but the Ventureño taxon *tash* seems to include *Juncus balticus* and perhaps small individuals of *J. textilis*. For the most part, however, *J. textilis* comprised a separate Chumash category called *mekhme'y*, discussed in the next section.

Harrington's consultants described *tash* as a kind of *junco* that had slender but tough stems about one-sixteenth inch in diameter. It did not grow in bunches. The plants could get as tall as *mekhme'y*, four feet high or more. This species was found growing in sand near the mouth of the Ventura River and "in the *montes*" (Sp., woods).

The principal use of *tash* was in the foundation of coiled basketry. To start the basket, a core of shredded *mekhme'y* was wrapped and sewn with split strands of the same material. Then, as the base enlarged, the shredded material was gradually replaced by whole stems of *tash;* on the main body of the basket, the foundation consisted of three whole *tash* stems, or rods, sewn with split *mekhme'y*. This information from Harrington's consultants agrees with the construction of actual Chumash baskets studied by Dawson and Deetz (1965), although those authors characterized both the sewing material and the three-rod foundation as *J. textilis.* I am somewhat skeptical of this identification; the foundation rods seen in most Chumash baskets are usually about one-sixteenth inch in diameter, while it is quite unusual to find *Juncus textilis* stems as slender as that.

The Chumash made many types of coiled baskets in this way, although consultants specifically mentioned *tash* in connection with only two. Large storage baskets, or *kh'i'm* (V, I), had the core of *tash* sewn with *mekhme'y* described above. They were made in various sizes, some up to a yard across, and tightly woven so that seeds kept in them would not leak out. Acorns, chia *(Salvia columbariae)*, *islay (Prunus ilicifolia)*, and other seeds were stored in these baskets, either inside people's houses or in cave caches in the mountains. Tarred water bottle baskets were also said to be made by coiling over a *tash* foundation (Craig 1966:209–210), but all documented examples of Chumash water bottles are twined, not coiled. An asphaltum impression of a basket that may have been a coiled water bottle, found on San Miguel Island and dated at 4500 years before present, is a possible exception (Braje, Erlandson, and Timbrook 2005).

Juncaceae — Rush Family

Juncus textilis BUCHENAU — **INDIAN RUSH**

(B, I, V) mekhme'y — (Sp) Junco

(Cr) meqme'i

(V) syɨt (reddish base of stems only)

The Spanish term "*junco*" refers to all members of the genus *Juncus*, but most specifically to *J. textilis.* The Chumash category *mekhme'y* is clearly equivalent to this species (except that very small plants of *J. textilis* may be included in *tash;* see previous section). *Mekhme'y* stems were from four to six or seven feet long, and over one-quarter inch thick at the base (Blackburn 1963:144; Craig 1966:206). No other species grows to that size. According to Harrington's consultants, specific locations where *mekhme'y* could be found included the Hueneme area, the sand dunes west of the mouth of the Ventura River, and Happy Canyon, in the Santa Ynez Valley. A number of Chumash place names meaning "at the reeds" or "place of the reeds" incorporate the

THE WORLD IS A BASKET

Chumash women are renowned for their skill and artistry in basketmaking. As she works, the weaver presses a sharp bone awl through the coiled foundation of slender rush stems. Quickly, before the tiny hole closes up, she inserts the wrapping strand of split, dyed rush and pulls it tight. She has just a few more rows of sewing to complete a large winnowing tray. Surrounding her are other baskets for cooking, carrying water, keeping valuables—even wearing as a hat. Baskets play essential roles in every aspect of her life.

name of this plant, including the town of *Kamekhme'y*, near the Ventura River mouth (Applegate 1975b:29, 31, 36). These place names may suggest additional locations for *Juncus textilis* in the Santa Ynez Mountains and along the Santa Barbara coast.

Mekhme'y was most important as a basketry material, particularly for the sewing strands in coiling. According to information from Candelaria Valenzuela and other Harrington consultants, straight *junco* stems were preferred; ordinarily those that grew twisted or crooked would be avoided. The Chumash cut or pulled up the whole stems and prepared them in various ways according to the color desired.

To cure the fresh, green *junco* and turn it a light tan color, they built a fire atop a thick bed of cold ashes. When the ashes had heated sufficiently, they ran the stems one by one in and out through the lower ash layer, continuing this process until the stems had turned from green to white. Then they laid the treated stems in the sun to dry before splitting them into strands (Blackburn 1963:144). Recent experiments with this method have resulted in golden-colored *Juncus* strands, much like the color of finished Chumash baskets (Lillian Smith, personal communication 1980). The ash treatment was done only with freshly cut *mekhme'y*, not with material that had already been dried.

Next they cut off the tip and base of each dry, cured stalk so the remaining portion was about four feet long. They split the stalk in half lengthwise, starting the split with the thumbnail and running the hand down the length of the stalk. Then they split each half again, to make four sewing strands in all. If the strands were to be dyed, that was done before any further cleaning (Craig 1966:206–207).

The Chumash had several ways of coloring the *mekhme'y* black. Candelaria Valenzuela demonstrated one, which was to roll the strands—which had been split but not cleaned—into a coiled bundle about a foot in diameter, tie it in several places with string, and bury it in black mud where there was considerable humus for a period of two to four weeks or until the strands were dark enough (Blackburn 1963:144–145; Craig 1966:208–209). When she dug up the bundle, she washed off the mud and allowed the strands to dry. A plant called *ya'i* (tentatively identified as *Lotus scoparius*) was burned for smoke in which to further blacken the strands (Heizer 1970:64).

Fernando Librado also described a method for blackening the *junco* by burying it. They would dig a trench and build a fire in the bottom. When the fire burned low, they put down a layer of the green *escoba* weed called *ya'i* and covered it with dirt. Then they laid on a bundle of split *mekhme'y* stems five inches in diameter, then more *escoba* weed, and then covered the hole with dirt and allowed it to remain for eight days. It is unclear whether this was a variation on Candelaria's method of mud burial or if it was a separate process.

The Chumash also dyed *Juncus* black with "ironwood bark" boiled in water and, historically, with a piece of iron added to pounded acorns placed in an *olla* filled with water. In both cases the split *junco* strands with the pith still on were soaked in the liquid. Though these methods were faster, they weakened the fibers and most people preferred the mud burial technique (Blackburn 1963:144; Craig 1966:208). It is unclear whether this "ironwood" was island ironwood (*Lyonothamnus floribundus*) or mountain-mahogany (*Cercocarpus betuloides*), which was also sometimes called ironwood.

Merrill (1923:221) included the Chumash in a list of southern California peoples who made a black basketry dye from "*Suaeda suffrutescens*, Sea-blight" (*sic:* sea blite, perhaps bush seepweed, *Suaeda moquinii*), attributing this information to Barrows (1900:43). Barrows, however, was describing Cahuilla basketry, not Chumash. Use of *Suaeda* by the Chumash does not seem to be confirmed in Harrington's notes, unless the unidentified plant called *ya'i* is *Suaeda;* but *Suaeda* does not seem to match the description and other uses of *ya'i.*

Consultants described other dyed colors that have not been identified in extant Chumash baskets. Two kinds of roots were said to be used for dyeing *junco* yellow. One of these was *cañagria (Rumex hymenosepalus);* the other may have been either leather root (*Hoita*, formerly *Psoralea*, *macrostachya*) or Durango root (*Datisca glomerata*). These roots were mashed and put in water, and the *junco* strips were allowed to soak in the liquid for several days. Fernando Librado mentioned baskets having red and black "paintings," though he may have meant woven-in designs (Craig 1967:114). Candelaria Valenzuela said a basket made by Juana Morales had blue designs made with "American dyes" (Heizer 1970:64).

Not all of the colors seen in Chumash baskets resulted from dyeing. Some *Juncus*, particularly the base of plants growing in sandy places, was naturally red-orange in color (Craig 1966:209; Heizer 1970:64). From four to twelve inches long, this distinctive kind of *mekhme'y* was known by a different term, *syɨt* (V). Weavers cut the reddish portion off from the regular, tan-colored part of the stem and used it separately for design elements on their baskets or for a deeper, richer-colored background.

Some *mekhme'y* was naturally white at the base and could be further whitened by the treatment with hot ashes described above. Chumash baskets with very light background color may be made from *Juncus* treated in this way, and white outlines around black designs on a dark orange background also appear. Bright, pearly white is usually a different material, split shoots of three-leaved sumac (*Rhus trilobata*).

If the *junco* was to be used in its natural color, it was cleaned and trimmed right after splitting. Otherwise the strands were dyed first and then cleaned. To clean the split *junco*, the basketmaker removed the cottony inner pith by drawing each strand over the sharpened edge of a clamshell while applying pressure with her thumb. This process also trimmed the strands to an even width for use as sewing material (Craig 1966:207–208). Bundles of *junco* basketry material were always stored in a cool place.

The basketmaker began to weave a coiled basket using a foundation core of finely shredded *mekhme'y*, which was flexible enough to be bent into a tight spiral at the start. She wrapped this foundation and sewed the coils together with split strands of *mekhme'y*. For each stitch, she pierced a hole through the foundation of the previous row with a bone awl, then quickly wrapped the sewing strand over the projecting foundation bundle and inserted it before the hole closed up again. She kept the awl in her hand the whole time and did not set it down between stitches. As the basket base enlarged, she gradually substituted whole stems of *tash* (*Juncus balticus* or very fine *J. textilis*) for the shredded material, so that most of the basket was woven on a foundation of three slender *Juncus* rods. The sewing material throughout the whole basket was split strands of *mekhme'y*.

While working on a basket, the weaver always kept a basin of water close by so the *junco* could be wetted from time to time. This was necessary to prevent either the sewing strands or the foundation material from breaking when bent. They might also break if soaked too long. (Harrington's observations of Candelaria Valenzuela making coiled baskets were compiled by Steve Craig: see Craig 1966.)

Based on Harrington's notes and other sources, Hudson and Blackburn (1982–87) have described a broad spectrum of Chumash basket types. Consultants explicitly stated that *mekhme'y* was used in making particular kinds of baskets: hats, burden baskets, hoppers, winnowers, parching trays, storage baskets, cups, and bottles.

Women's basketry hats, *'epsu* (V), were always made of coiled *mekhme'y*. Fernando Librado once saw one made entirely of *syɨt*, the natural red base of the *Juncus* stems.

Conical burden baskets, *helek'* (V) or *'akhtakuy* (I), were said to be made of coiled *mekhme'y*. They were carried on the back with the tumpline, or forehead band, attached to a string tied across the mouth of the basket. An Ineseño consultant said these coiled burden baskets were sometimes decorated with woven-in designs of people, toads, flies, and the like. Zoomorphic and anthropomorphic figures—or design elements readily recognized as such—are not characteristic of most known Chumash baskets, however.

A basket hopper, specially made without a bottom, was placed on top of the stone mortar so that acorns or other items would not fly out while being pounded. To make sure the hopper would fit the mortar for which it was intended, Fernando Librado said, a piece of *junco* was tied into a loop a little smaller than the mouth of the mortar. They used this as a measure to begin weaving the basket.

Seeds were winnowed in a large, shallow basketry tray made of *mekhme'y*. Known variously as *yɨw*, *'ayuhat* (V), or *'ayakuy* (I), these winnowing trays continued to be used in mission times for cleaning wheat. Another type of tray, called a *yep* (B), was similar but coated with tar, or with a mixture of tar and sand. Seeds were toasted by tossing them with hot coals in the *yep;* olivella-shell beads were whitened in the same way.

The Chumash kept large globular baskets for storing seed foods—acorns, *islay*, chia, red maids, and, historically, beans—either in their houses or in caves in the mountains. These *kh'i'm* (B, I, V) were up to two or three feet in diameter and sometimes tarred on the outside. They were sewn with *mekhme'y* on a foundation of either *tash (Juncus balticus)* or deer grass (*Muhlenbergia rigens*). For storing valuables, the Chumash also made a smaller type of storage basket, *kh'omho* (V) or *'alats'ɨmɨmɨ* (I), which had a narrow neck and sometimes a lid.

Small drinking cups made of *mekhme'y* could be either decorated with designs or plain; some had tar or pine pitch smeared over the inside (Heizer 1970:67). The drinking cup was called *'akmila'ash* (V) or *'akupusha'ash* (B, I). Tar-lined water bottle baskets, *'ush'e'm* (V) or *'awa'q* (B), might be made of *mekhme'y*, and some consultants asserted these were of coiled weave. No such examples are known, however, and most Chumash agreed that these baskets were twined of *'esmu (Juncus acutus).*

With few exceptions, this information on basketry forms and construction provided by Harrington's consultants is consistent with Dawson and Deetz's (1965) study of extant Chumash baskets. The latter authors identify the coil foundation material as *Juncus textilis*, while the Harrington data indicate it was probably *J. balticus.* Perhaps microscopic or chemical analysis could resolve which materials and dyes the Chumash actually used.

Juncus textilis did have one or two other uses besides in basketmaking. During mission times, the Indians bound the long, whole stalks together in bundles to make brooms for sweeping indoors. To treat poison-oak rash, they burned *mekhme'y* stems and rubbed the ashes on the affected part. Tule was also used in this way (see *Scirpus*, *Toxicodendron*).

Other Indian peoples, particularly the southern California mission groups, made baskets with *Juncus textilis*. Among them were Kitanemuk, Gabrielino, Cahuilla, and Luiseño (Harrington 1942:23, Bean and Saubel 1972:80–81, Coville 1904:207). The Chumash, Gabrielino, and Fernandeño sewed coiled baskets almost exclusively with *J. textilis* (Dawson and Deetz 1965:207), while elsewhere more sewing was done with other materials, such as three-leaved sumac (*Rhus trilobata*). Mud-burial and sea-blite dye methods were also widely reported. Another dye for *Juncus* was described by a Mrs. Hunter, who told Harrington she had two baskets made by Teodora, a Gabrielino woman. The brownish-black designs were supposedly dyed with a plant that grew near San Gabriel in the rainy season and had foot-high, iris-like leaves and blue flowers like daisies (perhaps blue-eyed grass, *Sisyrinchium bellum*).

Cupressaceae — Cypress Family

Juniperus californica CARRIÈRE — **CALIFORNIA JUNIPER**

(B, I, V) mulus (tree, fruit) — (Sp) Guata

(O) t'pi'nɨ

The Spanish name *guata* is a loan word from the Luiseño language term *waa'at*, juniper (Harrington 1944:38). Camulos, a place in the Piru region of Ventura County, derived its name from the Chumash word for a juniper tree that grew there (Applegate 1974:194).

From the wood of the California juniper the Chumash made sinew-backed bows which always retained their shape, no matter how much they were bent. The stave was cut green, allowed to dry, then worked until it was even and smooth. Fernando Librado had heard that juniper was very pitchy and likely to rot, and that it was unsuitable for bows that were not backed with sinew. Actual examples of both sinew-backed and unbacked bows have been collected in Chumash territory (Hudson 1974:3).

The berrylike seed cones of the juniper were ground up into a bright yellow meal called *hukhminash* (B, I), which had a sweet but resinous flavor. Fernando Librado tasted it but had to spit it out—he said there were two Indian foods he could not stomach, whale and *mulus*. Most of Harrington's consultants had eaten it, but few had seen it prepared. Some thought it was made from the "budding sprout tips" (perhaps the very young pollen cones) rather than juniper berries. Sometimes a little water was added so the meal could be molded into

CALIFORNIA JUNIPER
Juniperus californica

cakes eight inches or more in diameter. The cakes were hard when dry, but still crumbly. Yokuts and Kitanemuk people brought them in specially made tule containers to sell in Ventura and Santa Barbara. Apparently the coastal Chumash primarily acquired these juniper cakes through trade.

In August 1806, the Zalvidéa expedition found the Indian inhabitants of *Casteque*, an interior Chumash village, "away at their fields of *guata*." They also encountered Indians staying at camps in the San Gabriel foothills or Antelope Valley (Kitanemuk or Serrano territory) to gather the *guata* (Cook 1960:247). A priest at San Gabriel Mission said that the sweet resin called "*guautta*" [*sic*] was gathered in the fall (Geiger and Meighan 1976:81).

Indians and Spanish settlers in the Chumash region used a medicinal decoction of juniper to treat rheumatism and genito-urinary disorders (Bard 1894:6). Harrington's consultants did not mention this remedy.

Many Indian groups in California and the Great Basin used *Juniperus* species for bow making, food, and medicine (Mead 2003:217–220; Ford 1975:207). Unlike the Chumash, the Kitanemuk prized juniper wood for bow making (Harrington 1986). The Ohlone regarded *J. californica* berries as poor food but made a medicinal decoction of the leaves to relieve pain and cause sweating (Bocek 1984:248). The Cahuilla ate the berries fresh or prepared into mush or cakes and also used them medicinally (Bean and Saubel 1972:81–82). The Tubatulabal made bows from *J. osteosperma* wood and gathered the fruit in May (Voegelin 1938:27).

Jesus Javro, a Gabrielino Indian, told Harrington that *guata* was always eaten by people who were going to dance the next day. The Gabrielino gathered the berries and dried them in the sun, then pounded them in a stone mortar at night when it was cool; in the warmth of the daytime the resin would soften and become too gummy. They put the pounded berries into a large basket with a little water to make a dough, adding more pounded berries and mashing them all together until the basket was full. They allowed this material to stand in the basket until it hardened into a solid loaf and then broke it into pieces to throw to the crowd assembled for the dance. The Indians scrambled to get it. Occasionally they boiled the fresh juniper berries, but usually they did not toast or cook them at all (Harrington 1986).

Keckiella cordifolia: See *Lonicera subspicata* var. *subspicata*

Brassicaceae — Mustard Family

Lepidium nitidum TORREY & A. GRAY — **PEPPERGRASS**

(B, I) khaksh — (Sp) Tapona

(V) 'iqma'y

The Chumash toasted the small, yellow, mustardlike seeds of peppergrass by tossing them in a basket with hot coals, then ground them and ate this pinole. To treat diarrhea and dysentery, they simmered a leaf or two of this plant in water and drank the decoction while it was still warm (Birabent n.d.). If a strong laxative, such as elderberry root, had been taken in medicinal treatment, one could counteract the effects by drinking peppergrass seeds pounded up in cold water. The Spanish name reflects this quality; *tapón* means "plug."

Poaceae — Grass Family

Leymus condensatus (C.PRESL.) A. LÖVE — **GIANT RYE**

(B) shtemelel — (Sp) Carrizo

(Cr, I, V) shakh

(O) tqmimu'

(P) shtemele

The Chumash applied the Spanish word "*carrizo*" to three large grasses: *Leymus* (formerly *Elymus*) *condensatus*, *Phragmites australis*, and the introduced species *Arundo donax*. They often distinguished between the latter two as "sugar *carrizo*" and "*carrizo de Castilla*," respectively. All three could be used interchangeably for certain purposes. Until mission times, *Leymus* was the only kind of *carrizo* that was locally available along the coast and on the islands.

Although some of Harrington's Chumash consultants claimed this species was too light for arrows, others said that giant rye was used like the other *carrizos* for making arrow shafts. They would dry the stems thoroughly before straightening them. Then they inserted a wooden foreshaft of *romerillo (Artemisia californica)* or toyon *(Heteromeles arbutifolia)* into the *carrizo* shaft, glued it in place with tar, and wrapped it with sinew. A stone point was optional. The fletching was three feathers. These arrows were reportedly used only for rabbits, birds, and other small game.

The Chumash sometimes thatched their houses with giant rye, particularly on the Channel Islands, where other kinds of thatching materials were scarce. Houses thatched or walled with this *carrizo* were specifically mentioned at the towns of *Kamekhme'y*, near Ventura, and Cieneguitas, near the present Hope Ranch, and at Sanja Cota in Santa Ynez. House design changed in post-mission times: some of these dwellings were oblong, or of brush plastered with adobe but with a *carrizo* thatched roof. By this time *Arundo* had become commonly available. Though tule mats were more often used, the Chumash sometimes made their house doors of *carrizo*. The door was two inches thick, woven with *tok (Apocynum cannabinum)* rope, and with one horizontal brace (Hudson and Blackburn 1983:330).

The hollow, canelike stems of giant rye proved useful in several ways. For smoking tobacco, tubular cane pipes functioned much like cigarettes; the end of the tube was plugged with three slender sticks to keep the tobacco from falling into the mouth. The tobacco, either alone or in a mixture with burned shells, was stored and carried in small tubular containers of *carrizo*. Men frequently wore these as ornaments in the pierced nasal septum or the earlobe, or over the top of the ear (Harrington 1942:16, 27). *Carrizo* stems inserted into the holes in newly pierced ears kept them from closing while they healed. A *carrizo* stem could also be used as a straw. Harrington reported that Chumash people used carrizo tubes to suck water through a plugged chink in a rock at a perennial spring (Harrington 1935:84).

A five-inch section of dry *Leymus* stem, split lengthwise and the edges scraped, made an extremely sharp cane knife. These knives were usually pointed at the end and not hafted into a handle. Employed with a sawing motion, this *carrizo* knife worked well for gutting animals and for cutting meat and fish. The Chumash also cut umbilical cords of newborn babies with the *carrizo* knife, preferring it to a flint knife for this purpose because the cut would heal faster when cut with *carrizo*.

The Chumash made paintbrushes from raccoon or badger tails with *carrizo* handles. Leaving only a small tuft at the very end, they scraped all the hair off the last inch and a half of the tail and inserted the bare part, with the bones still inside, into the end of a *carrizo* tube about six inches long, gluing it in place with pitch and tying it tightly. They used these brushes for painting anything, but especially for applying body paint for dances.

Carrizo counter sticks kept the score in women's walnut-shell dice games (see *Juglans californica*). The scorekeeper had a certain number of *carrizo* sticks about eight inches long. If a player scored, the scorekeeper tossed her a *carrizo* stick and she threw again. The final winner was the one who had accumulated all the *carrizo* sticks.

Some consultants thought that giant rye was too thin and light to be good for whistles or flutes; probably *Phragmites*, and later *Arundo*, was used instead. Others mentioned baskets made of *carrizo* shoots and aprons made of *carrizo* cordage, but these uses of *Leymus* have not been substantiated.

For medicinal purposes, people would collect the new shoots of giant rye just as they began to emerge in March. One treatment they mentioned was to boil half a handful of *carrizo* shoots in a quart of water and drink it as a sure cure for gonorrhea. By May, the plants were too far advanced to be employed in this way.

The Chumash name for May was *hesiq'momoy 'an maishakhuch*, "month when *carrizo* is abundant." A person born in May would have knowledge of fundamental things for the good of humankind, such as medicine (Blackburn 1975:101). The phrase *shtalhɨw 'ishakh*, "son of a *carrizo* clump," was applied to a child born to any unmarried woman. These Chumash phrases refer to *shakh*, giant rye (*Leymus condensatus*), not to *carrizo* grass (*Phragmites*), which had different names in Chumash languages.

Among the Cahuilla, *Leymus condensatus* served two important functions: arrow shafts and house thatching. The Luiseño also used giant rye in arrow-making (Bean and Saubel 1972:69). Some California peoples ate the seeds of this plant (Heizer and Elsasser 1980:245; Mead 2003:153–154), but there is no record of the Chumash having done so.

Apiaceae — Carrot Family

Lomatium californicum (TORREY & A. GRAY) MATHIAS & CONSTANCE — CHUCHUPATE

(B, I) pa'
(V) chpa'
(Sp) Chuchupate

Throughout California, the Great Basin, the Southwest, and Mexico, the name *chuchupate* refers to certain members of the Apiaceae, or carrot family. Most have aromatic roots that were used medicinally and in rituals. In much of western North America, the principal species of *chuchupate* was *Ligusticum porteri* (also called *oshá*), whereas *Angelica* and *Lomatium* spp. tended to be more important in California (Ford 1975:183, 253–255; Jepson 1936:633; Mead 1972:15, 120–122; Strike

1994:13–14, 84–85). It also appears that a plant often known by the common name "angelica," used ritually by several northern and central California groups, may in fact be *Lomatium californicum* (Lawrence Dawson, personal communication 1990). Among the Chumash, in any case, *chuchupate* was clearly *Lomatium californicum*, an uncommon plant of interior mountain chaparral. A sample of this same species, labeled "*chuchupate:* root, flower, leaves," appears in a photo taken by Harrington among the Yokuts in 1916–1917 (1994: reel 4, fr. 0187)

The principal Chumash use of *Lomatium californicum* was in the control of rattlesnakes, which was accomplished in various ways. One could carry a piece of *chuchupate* root in the clothing or wear it as a talisman for protection against rattlesnakes; the snake would rattle before the person got close and it would not bite. One could also chew the *chuchupate* root and then throw that onto the snake to stupefy it.

Professional Rattlesnake Dancers went out to capture snakes two or three days before they were to dance. They said it was best to get the snakes in the daytime during a period of either full moon or new moon. To catch a snake, a man chewed *chuchupate* root and spat on his hands. He held his hand out toward the rattlesnake from the windward side. The snake would go to sleep and the man would pick it up and put it in a bag. Another method was to spit the chewed *chuchupate* root into the bag and place it on the ground, and the snake would crawl inside (Hudson et al. 1977:87).

An even more dramatic way of capturing rattlesnakes was also reported. *Chuchupate* root was pounded up and mixed with a little water. One man held the snake down with a forked stick behind its head, placed his foot on the middle of its body, and with two little sticks propped its mouth open. The other man took a big mouthful of the *chuchupate* liquid and, through a hollow *carrizo* stem, squirted it directly into the snake's mouth. Some snakes were said to be able to stand a quantity of this, but others soon became drunk and tried to roll over on their backs. At this point they were easily picked up.

The snakes were kept for three days, and on the morning of the fiesta the *chuchupate* liquid was squirted into their mouths again. There was no song for the Rattlesnake Dance; the man just danced, moving his feet gently. None of the onlookers were allowed to get close to the snake (Hudson et al. 1977:87–88). This dance may have been intended to protect the entire community from snakebite (Walker and Hudson 1993:100).

The Chumash regarded *Lomatium californicum* as an important medicinal plant, particularly to treat pain and to improve digestion. For rheumatism they drank a tea of the root, or chewed it and rubbed the paste on the site of any kind of pain. They also chewed the root as a tonic, and it was said to be useful in the treatment of flatulence, headache, and neuralgia (Bard 1894:6). The root was boiled into a decoction or, historically, minced and soaked in brandy and drunk for stomach ailments (Birabent n.d.). Some of Harrington's consultants said that *chuchupate* root was "eaten," but they may have been referring to its being chewed for medicinal or magical purposes.

Early in mission times the Spanish adopted many medicinal uses of this plant from the Chumash, including inhaling the scent to cure headache and chewing the root for stomach trouble. Chumash people sold pieces of *chuchupate* root to the presidio soldiers for high prices, claiming that it was very rare (Longinos Martínez 1961:45–46; Geiger and Meighan 1976:73). A missionary recommended the use of *chuchupate* to treat facial paralysis, pleurisy, and spleen disorders (Garriga 1978:34, 35, 38).

Even in recent decades Chumash and Hispanic people have continued to use *chuchupate* in various medicinal ways, chewing the root or taking a decoction for cough, sore throat, nausea, upset stomach, diarrhea, constipation, toothache, headache, and other complaints. Some are known to carry pieces of the root in a pocket in case it might be needed (Gardner 1965:298; Brand and Townsend 1978:25).

Chemical analysis of essential oils of this species found high concentrations of compounds with sedative and muscle relaxant effects (Beauchamp et al., 1993).

Lonicera subspicata var. *denudata:* See *Symphoricarpos*

SANTA BARBARA HONEYSUCKLE
Lonicera subspicata var. *subspicata*

Caprifoliaceae — Honeysuckle Family

Lonicera subspicata HOOK & ARN. var. *subspicata* — **SANTA BARBARA HONEYSUCKLE**

(Sp) Moronel

Scrophulariaceae — Figwort Family

Keckiella cordifolia (BENTH.) STRAW — **CLIMBING PENSTEMON**

(B) tenech

(Sp) Moronel

These two plants are unrelated but similar in being woody and vine-like, clambering over other shrubs. The Chumash more or less lumped them into one category, according to plant specimens collected by Lucrecia García in 1929. She labeled *Keckiella* as *moronel*, and this variety of *Lonicera subspicata* as "another kind of *moronel.*" A separate category lumped the other honeysuckle variety, *Lonicera subspicata* var. *denudata* (formerly *L. johnstonii*), together with snowberry (*Symphoricarpos mollis*).

Moronel leaves were boiled in water and the decoction drunk as a treatment for colds (Birabent n.d.). On a tag accompanying the *Lonicera* specimen, Harrington noted that a tea of that plant was drunk for runny nose. Sores and wounds were washed with a tea of *moronel* leaves, then the fresh leaves were applied as a poultice (Bard 1894:7; Garriga 1978:43).

Other peoples used *Lonicera* species in similar medicinal ways. The Paipai of Baja California also bathed bruises and wounds with tea of *moronel* (*L. johnstonii*, now *L. subspicata* var. *denudata*), and the Yuki washed sore eyes with a tea of *L. interrupta* (Ford 1975:243; Mead 2003:240).

Fabaceae — Pea Family

Lotus scoparius (NUTT.) OTTLEY (?)

CALIFORNIA BROOM, DEERWEED

(B) ya'i

(V) yai

(Sp) Escoba de Horno

The identification of the plant known to Harrington's consultants as *ya'i* or *escoba de horno* (oven-broom) is based largely on the fact that, in the early days, brooms were made from *Lotus scoparius* at Santa Barbara and elsewhere in California (Smith 1998:245). Two specimens of this plant in the Harrington collection, however, were labeled "wild alfalfa" by Lucrecia García. *Escoba amarilla* was a different plant (see *Hemizonia fasciculata*).

The Ventureño Chumash made brooms of *escoba de horno* and several other kinds of plants (see also *Hemizonia*, *Solidago*, *Symphoricarpos*) for rough sweeping outdoors or for cleaning out the mission ovens. The branches alone might be used for this purpose, but often a stick was attached as a handle.

This same plant was also used in coloring *Juncus* black for baskets. After the split *Juncus* stems had been buried in black mud for a few weeks, they were dug up, washed off, and allowed to dry. Then *ya'i* was burned to make a smoke in which the strands were further blackened (Heizer 1970:64). A fire was built in a trench and after it burned down to coals, a layer of green *escoba* weed was put on and covered with dirt, a bundle of split *Juncus* was laid on, and more *escoba* weed placed over it. Finally the hole was covered with dirt and allowed to remain for eight days.

Ya'i was the only plant the Chumash used for thatching sweathouses, said Fernando Librado, because it was not flammable. They placed it over a framework of willow poles.

Fabaceae — Pea Family

Lupinus spp. — **LUPINE**

(B) wala'laq', qlaha' — (Sp) Perro, Perrito

(V) qlahaw'

There are nine lupine specimens in Harrington's Chumash materials at the National Anthropological Archives, representing several different species, and Lucrecia García labeled them all as *wala'laq'*. In Harrington's notes, however, the names *qlaha'* (B) and *qlahau'* (V) are associated with lupine. In one passage, Candelaria Valenzuela describes preparation of lupine seeds for food (Craig 1967:100), but that process was exactly the same as the one followed for wild cherry (*Prunus ilicifolia*). No other consultants reported use of lupine as food. I believe there was some confusion in these notes and that the Chumash did not eat seeds of lupine. Some species are considered toxic, especially when eaten in quantity (Fuller and McClintock 1986:161–164).

Rosaceae — Rose Family

Lyonothamnus floribundus A. GRAY — **SANTA CRUZ ISLAND IRONWOOD**

subsp. *asplenifolius* (E. GREENE) RAVEN

(Cr) wɨlɨ — (Sp) Palo Fierro

(V) wɨ'lɨ

The common names "ironwood" and "*palo fierro*" have been applied to a number of unrelated plants that share the characteristic of having very dense wood, including the local mountain-mahogany (see *Cercocarpus*). Desert ironwood (*Olneya tesota*), well known as a source of hard wood for Seri Indian carvings, is not found near Chumash territory. The uses listed here are those attributed to Santa Cruz Island ironwood, also called "fern-leaved Catalina ironwood." This plant is indigenous only to Santa Cruz, Santa Rosa, and San Clemente Islands, and a related subspecies is found on Santa Catalina.

Ironwood was the principal wood the Chumash used for making harpoons. These were eight feet long and had a bone point to which the line was tied. When they harpooned a sea mammal or fish, the point stuck fast, while the shaft came loose and remained floating on the water so it could be retrieved later. It is unclear from Harrington's notes whether the spear for barracuda was the same as a harpoon; the barracuda spear was described as being eight feet long and made of two pieces of ironwood joined end to end.

The Chumash also preferred ironwood for making the shafts of canoe paddles because of its strength. They cut pieces six to eight or even twelve feet long, allowed them to dry, and then worked them until they were smooth, tapering both ends to a point for attachment of the paddle blades. Sometimes it was necessary to straighten the paddle shaft with fire.

Abalone meat was removed from the shell using a tool made either of seal shin bones or, on the islands, a stick of ironwood a foot long and shaped like a shovel. This implement was different from the pry bar with which the abalones were collected.

For fighting, the Chumash made wooden knives of ironwood about twenty inches long and pointed on both ends. There was a groove at the center for a handgrip, and the knife was painted with red ochre. One gripped it at the center and pierced with both forward and backward thrusts.

Archaeological excavations on Santa Rosa Island uncovered well-preserved remains of a Chumash house that had a structure of seven split ironwood posts and four whale ribs. It was covered with seagrass matting that was tied on and held in place with large, flat stones (Rogers 1929:331–332). A number of similar house remains have been found, but in most cases what remained of the wood was not identified. Since the willow (*Salix*) and tule (*Scirpus*) that were usually used to construct mainland houses were scarce, the island Chumash substituted local resources. Fernando Librado indicated that some people may also have shredded ironwood bark to make women's fiber aprons.

According to Candelaria Valenzuela, weavers would sometimes boil ironwood bark into a black dye for basketry material. The *Juncus* stems were split but not cleaned, then coiled up and allowed to soak in this solution for a week or so. Most people did not favor this method as it tended to weaken the fibers, although they said it was good for dyeing buckskins. It seems curious that Candelaria Valenzuela was aware of this plant. Perhaps the bark was formerly traded to the mainland, and by her day the tree may have been introduced into cultivation as an ornamental; or she may have been referring to a different plant with the same common name.

Malvaceae — Mallow Family

Malacothamnus fasciculatus (TORREY & A. GRAY) E. GREENE — **CHAPARRAL MALLOW**

(B) khman — (Sp) Malva

Fernando Librado told Harrington that the Chumash made cordage from the "bark" of this shrub (probably actually the stem fibers under the bark), in much the same way they did from Indian-hemp (*Apocynum*) or milkweed (*Asclepias*). They wove this string to make sacks they used for gathering brodiaea bulbs and the like. Other consultants did not confirm this use and, although it is in the same plant family as jute, *Malacothamnus* does not appear to have been an important fiber plant for the Chumash. Specimens were labeled with Barbareño names.

The Luiseño made a medicinal decoction of the leaves of this species for use as an emetic (Sparkman 1908:231). Generally, members of the related genus *Malva* seem to have been of greater medicinal importance than *Malacothamnus*.

Anacardiaceae — Sumac Family

Malosma laurina (NUTT.) ABRAMS — **LAUREL SUMAC**

(B, I, V) walqaqsh — (Sp) Mangle Mayor

The Barbareño and Ineseño Chumash included *Malosma* (formerly *Rhus*) *laurina* in the same folk taxonomic category as lemonadeberry and sugar bush (see also *Rhus integrifolia* and *R. ovata*). The Ventureño divided that category and recognized a larger and a smaller kind, with laurel sumac comprising the "larger" form, or *mangle mayor* (Sp). The Spanish common name "*mangle*" refers to the similar looking mangroves found in tropical wetlands.

Chumash people pounded the berries of laurel sumac, dried them in the sun, and ate them raw. Fernando Librado told Harrington that the island Indians winnowed the chaff from the dry pulverized *mangle* fruit in an abalone shell. He also said that the Chumash scraped the bark from laurel sumac root, boiled it in water, and drank it for dysentery.

Malvaceae — Mallow Family

Malva nicaeensis ALL. (Introduced) — **BULL MALLOW**

M. parviflora L. (Introduced) — **CHEESEWEED**

(B, I) mal — (Sp) Malva

(O) temala

(V) malwash

Both these species are weeds introduced from Europe, but they have been used by many New World peoples for food and medicine. The common names in Chumash were derived from the Spanish name "*malva*." The Chumash ate *malva* seeds or fruit, which were especially popular with children.

Several of Harrington's consultants commented that *malva* was a remedy of the white people. They either knew or themselves practiced a number of these medicinal uses for both roots and leaves. Luisa Ygnacio said the whites would pound the roots from an old *malva* plant and boil this in water to make a tea to drink to treat fevers; María Solares prepared this tea to drink for the blood. Other sources mention boiling the whole plant in water as a warm bath for inflammation in horses, as well as boiling the root in milk for colds, cough, fever, and "fever in the stomach" (Birabent n.d.; Gardner 1965:299).

Some people ground up *malva* leaves to use as a medicine for sore, swollen knees, according to Juan Justo. Luisa Ygnacio said the Spanish boiled the leaves and applied them as a poultice for eye trouble. More recently, people have boiled the leaves into a wash or mixed them with lard for use as an ointment for infected sores; alleviated dandruff with a *malva* infusion; and treated "*empacho*," a kind of stomach trouble, with *malva* mixed with rose petals and other herbs (Weyrauch 1982:11). Juanita Centeno said to fry *malva* leaves lightly in oil and apply them to the chest for chest colds (personal communication 1978).

Other California Indians used *Malva* species in similar ways. The Cahuilla ate the fruits (Bean and Saubel 1972:88). The Ohlone made hot poultices of the leaves for stomach or head pain, and a decoction of the root for fevers or as a hair rinse, among other uses (Bocek 1984:250). The Miwok also applied *Malva* poultices for sores and swellings (Mead 2003:253).

Spanish-speaking people in Mexico and the Southwest reportedly use *Malva parviflora* and related species to treat fever, stomach trouble, boils, and bruises, and as a hair wash, among many other uses (Ford 1975:229–231). The Chumash probably adopted these remedies, along with the *Malva* plants, from outsiders.

Cucurbitaceae — Gourd Family

Marah macrocarpus (E. GREENE) E. GREENE **WILD CUCUMBER, MAN ROOT**

(B) molo'wot' — (Sp) Chilicote, Chilecote

(I) cha'

(V) 'anmakhwaka'y

Both children and adults pierced *chilicote* seeds and strung them into necklaces to wear just for fun. Typically these necklaces had three strands: one of yellow or whitish seeds, the center one of red, and the third strand of black seeds. Young boys also used the colored *chilicote* seeds as gaming pieces, much like marbles (Hudson and Blackburn 1986:371–372).

Chilicote was most important for its use in medicine. A paste made from the seeds could be stored for later use; Luisa Ygnacio's mother kept two abalone shells full of it, one inverted on top of the other. To make the *chilicote* paste, one would toast the seeds on hot coals until they were at the point of burning and had become very black, then remove the kernels from the shells and pound them to a doughy consistency. A little of the *chilicote* paste was put in hot water to make a medicine taken by pregnant women for the blood, to ensure that the child would be born all right. This same drink was given to babies or young children as a purgative. The liquid was also sprinkled on a boil or tumor to make it go away.

A special paint for curing someone who was seriously ill was also made from *chilicote* kernels that had been toasted until black and greasy. The curer painted lines or dots in particular places on the patient's body, depending on the nature of the illness. After singing, the curer then rubbed the patient's body—arms, torso, and legs—in certain ways. This removed the sickness so that it left the body. Soon

WILD CUCUMBER
Marah macrocarpus

after the patient recovered, he or she was given *Datura* to drink so that harmony would be restored.

An 1814 document from Mission San Fernando, where there were some Chumash people living, states that the Indians mixed *chilicote* (called "*yjaihix*" in the local native language) with powder from a stone called *pafa* or *paheasa*. They employed this mixture to reduce inflammations, heal wounds, remove cataracts, induce menstruation, and cure urinary disorders; they also boiled it and drank it to induce sweating (Kroeber 1908:15; Geiger and Meighan 1976:73).

The Chumash used *chilicote* as a hairdressing. The oil from the toasted seeds was thought to promote hair growth, according to Bard (1894:7), but another source asserts that it was the roots that were used in this way (Garriga 1978:30).

Chilicote was used as a special container for *'ayip*, a poisonous compound made with the mineral alum that the Chumash considered to be very supernaturally powerful and potentially dangerous. One could only carry *'ayip* with safety if it were placed inside a *chilicote* shell. The oil in the spine-covered fruit was apparently thought to seal in the harmful qualities.

The noted rock-art expert Campbell Grant suggested that the Chumash extracted the oil from *Marah* seeds for use as a binder for pigments such as hematite or graphite (Grant 1965:85–86). Since the kernels make an intense black, greasy form of charcoal when toasted, however, it is possible that the burned kernels themselves constituted the black pigment in rock art. This use was not mentioned by Harrington's Chumash consultants, who had very little to say about rock painting.

The Luiseño made both black and red paint used in rock art from the ground seed kernels of *chilicote* (formerly *Echinocystis macrocarpa*, now *Marah macrocarpus*). For black pigment, they charred the seeds and pulverized them with manganese oxide. For red, they toasted the seeds more lightly to extract the oil, then pounded them up fine and mixed in the pitch from bigcone Douglas-fir (*Pseudotsuga macrocarpa*) and a red pigment derived from an iron-rich bacterial scum found in certain springs (Sparkman 1908:209–210; Harrington 1933:142). Closer to the Chumash, the Yokuts were said to use *chilicote* oil in making paint (Latta 1977:594), but the documentation is ambiguous as to whether it was used as binder or as pigment.

There is greater agreement in the literature regarding the use of seeds and roots of *Marah* spp. by many California peoples to treat sores and as a purgative (Strike 1994:89–90; Mead 2003:254–255). Interestingly, the Yuki employed these seeds to commit suicide or to poison others (Chesnut 1902:390–391). The Chumash do not seem to have utilized this plant as a fish poison, a practice reported for several other groups.

CHUMASH ROCK PAINTING

In a remote mountain cave, a Chumash member of the esoteric 'antap *cult prepares red, black, and white pigments. He grinds each one to a fine powder, then mixes it with water to make a bright, long-lasting paint. Another* 'antap *member quietly chants prayers as he seeks to influence supernatural forces by painting their images on the cave wall. Their help is essential to success in all activities throughout a person's life.*

Lamiaceae | Mint Family
Marrubium vulgare L. (Introduced) **HOREHOUND**
(no Chumash name known) (Sp) Marrubio

The Franciscan friars introduced this plant as a medicinal herb in the late eighteenth century (Jepson 1943:397–398). At the missions, horehound was made into a tea drunk to relieve asthma, colds, and jaundice (Garriga 1978:18, 23, 32). Harrington's consultants barely mentioned this plant—only to say it was not medicinal—and it does not seem to have been important to the Chumash in the early days. In the mid-twentieth century, Santa Ynez Chumash stated that horehound-leaf tea was formerly drunk to induce abortion (Gardner 1965: 296–297). People have continued to take this tea for cough, colds, and sore throat (Weyrauch 1982:10). The latter uses are consistent with those reported among some other California Indians (Mead 2003:256) and Hispanic peoples in the Southwest (Ford 1975: 237–238).

Mentha spp. (Mint): See *Satureja*

Poaceae | Grass Family
Muhlenbergia rigens (BENTH.) A. HITCHC. **DEER GRASS**
(no Chumash name known) (no Spanish name recorded)

A distinctive feature of most Chumash baskets is their coil foundation of three *Juncus* rods. A few Chumash baskets, particularly those found in dry caves in the interior backcountry, are made with a grass bundle foundation. Quantities of prepared grass stalks have also been found in cave caches within the Chumash region. These are the flower stalks of this robust perennial bunch grass. Although its occasional use in Chumash basketry is documented by examples in museum collections, deer grass was not mentioned by any of Harrington's Chumash consultants.

Deer grass was much more commonly used as basketry foundation material among other groups in central and southern California than it was among the Chumash (Mead 2003:268–269). Many of these other groups burned the plants periodically to improve quality and increase production of the flower stalks (Anderson 1996).

Solanaceae | Nightshade Family

Nicotiana attenuata TORREY | **COYOTE TOBACCO**

Nicotiana clevelandii A. GRAY | (No English name known)

Nicotiana quadrivalvis PURSH | **INDIAN TOBACCO**

(B, I, V) show | (Sp) Pespibata, Tabaco

(O) stu'yi'

Important Note: All species of *Nicotiana* are toxic. Smoking these leaves may cause nicotine poisoning with unpleasant effects; ingestion is more dangerous and may even be fatal (Fuller and McClintock 1986:241–242).

The only native tobacco species known from the immediate Chumash coastal area and Channel Islands is *Nicotiana clevelandii*, which an early botanist observed growing only on former Indian village sites (Rothrock 1878:48). Much of its former coastal habitat has been affected by development. Luisa Ygnacio told Harrington that there used to be tobacco growing at Las Positas, in Santa Barbara, and in the 1880s Henshaw also said it grew in abundance around Santa Barbara (Henshaw 1955:152).

Within Chumash territory, the other two native tobacco species, *Nicotiana attenuata* and *N. quadrivalvis*, are much more commonly found in the interior mountains and valleys. Fernando Librado said that tobacco grew wild in the Tejon country. Candelaria Valenzuela also mentioned the Tejon location, adding that floodwaters of Piru Creek often carry the seed downstream into the Santa Clara River, and that the plant had also been seen growing near Saticoy (Blackburn 1963:144). A village in Piru Canyon, Ventura County, was called *Kashowshow*, "much tobacco" (Applegate 1975b:32).

Tree tobacco, *Nicotiana glauca*, was introduced from its native South America during mission times and has become widespread along roadsides. Shown a sample of this species by Arthur Harrington in 1925, Juan Justo gave the same Barbareño Chumash name as the native tobacco and said that it was ground up and eaten. Other consultants more familiar with traditional Chumash plant uses did not mention use of tree tobacco.

The term *show* refers primarily to the tobacco plant. Tobacco that had been prepared for use was known as *pespibata*, a word the Chumash apparently hispanicized from the Fernandeño term *pispivata*.

Considerable linguistic borrowing was associated with tobacco use among California and Great Basin Indian peoples. Similar tobacco preparation methods and customs are found in much of central and southern California. The Uto-Aztecans (also called Shoshoneans) in this region, and perhaps Chumash peoples as well, may have acquired a particular set of tobacco names from the Yokuts word *so:Gon* or *shookun* along with the species *Nicotiana quadrivalvis* (formerly *N. bigelovii*). Another set of Shoshonean terms, including Luiseño and Cahuilla *pivat* and Fernandeño *pispivata*, is associated in some groups with *N. attenuata* (Zigmond 1941:229–240). Some California peoples preferred *N. quadrivalvis* over *N. attenuata* (Zigmond 1941:244), but it is uncertain whether this was true of the Chumash.

Tobacco was the only recreational drug used by the Chumash, though it also had ritual and medicinal applications. For social use, tobacco was most often eaten after being mixed with lime from burned shell. Smoking, which utilized pure, unmixed tobacco, was most often done for curing purposes and seldom for recreation. The following information on Chumash tobacco use has been gleaned primarily from Harrington's notes. Although Candelaria Valenzuela said that the tobacco smoked by the Indians came from Tejon already prepared in packages (Blackburn 1963:144), it is clear the Chumash were preparing their own tobacco as well.

Some other Indian peoples sowed tobacco seed, weeded and cultivated the plants, or pruned them to encourage larger leaf size, but apparently the Chumash only used wild tobacco in its natural form. They picked the leaves from the plants in springtime and spread them out on the ground or on rocks in the sunshine to dry, turning them often so they would not get moldy. The leaves could also be dried on heated stones (Henshaw 1955:152). Even after drying, the leaves remained somewhat gummy.

To prepare the tobacco mixture, several mussel shells were placed in a fire and fanned with a basketry tray. Oyster or abalone shells might also be used, but mussel seems to have been preferred. After they had turned first red, then entirely white from the heat, the shells were removed with a stick and powdered in a small stone mortar. A small amount of dry tobacco was then added and the combination pounded again, mixing in a little water. The resulting *pespibata* was molded with the hands into cakes of various shapes—round balls five inches in diameter, lumps the size and shape of a small potato, or long ones, or like a brick—and bluish gray in color because they contained so much ash. Though Fernando Librado told Harrington the purpose of the shell ash was to "moderate" the tobacco, most consultants agreed that it made the tobacco much stronger. The old-timers liked to make tobacco stronger, they said, for it was not strong enough the way it grew.

The cakes of *pespibata* retained their shape when dry. They could be stored for a long time and were an important trade item. Tobacco, either plain or mixed with lime, was also kept in tubes made from segments of *carrizo* grass stems or elderberry wood with the pith removed. Stoppers on either end kept the tobacco or *pespibata* from falling out. Men carried these tobacco tubes behind their ears or even in their pierced earlobes or nasal septa. Historically, cattle horns were used for carrying tobacco. *Pespibata* was also kept in the house, among a person's most treasured things.

The *pespibata* mixture was often consumed as a drink, which was made by adding more cold water at the time of initial preparation or by pounding part of a dried cake in a small mortar with a quantity of water. This liquid was drunk directly from the mortar. Tobacco drinking was done especially in the late afternoon before supper, which was the main meal of the day. After drinking the liquid, one went outside, induced vomiting by gagging with the fingers, and returned feeling light and refreshed, with a good appetite for supper.

One consultant said that Chumash men formerly had many meetings at which they would talk informally and eat *pespibata*, much as people serve coffee or chocolate nowadays. About once a week several men who were learned in Chumash traditional ways would get together for a *pespibata* preparing ceremony, called *'alutapɨsh*. This activity, always done at night, seems to have been distinct from drinking tobacco before dinner. The shells were burned and ground together with tobacco as described above. The man who prepared the mixture would lick the pestle, for the tobacco that stuck to it was particularly tasty. He then gave a handful to the first man on his left. This man chewed it three or four times and passed it to the next man, who chewed it and in turn passed it along to the next, until all had partaken. They referred to this as "eating" *pespibata* because it was chewed and swallowed. Another method was to suck tobacco liquid from the fingers or from a sponge of tule pith dipped into the mortar. With either method, immediately after ingesting the *pespibata* each man went outside to vomit, then came back inside and sat down in a row again. One by one, the men presented their tobacco tubes to the *'alushwil*, or distributor, who filled them with a quantity of *pespibata* sufficient to last each man for the coming week. That way he could eat a little whenever he wished. In return for his portion each man gave something, perhaps money or a handkerchief, but if he had nothing he was not required to pay.

The *pespibata* tastes so good, the Chumash said, that it dissolves in the mouth like breath. Eating *pespibata* had an intoxicating effect that was considered pleasurable. Users were described as acting drunk, being in a semi-stupor, or feeling sleepy. In mission times the priests tried to discourage the practice, believing it led boys to drink beer and wine later on. At Mission Santa Inés the Indian *alcaldes* would look for *pespibata* indulgers and punish them by whipping.

Eating *pespibata* or sucking the liquid from the tule sponge was sometimes done as a test of strength. At festivals, among all the other games and contests, young men might stand around and pass the *pespibata* from one to the next to see who could stand the most rounds. It was hotter than the strongest chile, and it made them dizzy. The effect was immediate, not slow like wine; the man fell over backwards before he had time to swallow. Some valiant ones could endure three sucks before they staggered down. There was often heavy betting on these contests, and the wagering helped to raise money for the captain to defray his expenses.

For recreational purposes, tobacco was occasionally smoked in cigarette-like hollow *carrizo* tubes four or five inches long, cut from between two joints or nodes of the plant's stem. A man would make several of these at a time and fill them with tobacco when they were to be used. He lit the "cigarette," smoked two or three puffs, and stuck it back behind his ear. It went out if not smoked. Then in a while he would take two or three more puffs. He might trim off the end of the tube after the tobacco had burned down an inch or so, to make it draw better. When the tube had been smoked down all the way, the stub was always thrown in the fire, never just discarded. Tobacco leaves intended for smoking were, like *pespibata*, dried and pounded in mortars, but no shell ash or water were added because they would prevent the tobacco from burning well.

The Chumash had no cigars and never wrapped tobacco leaves into a cigar. Hollow elderberry stems may have been fashioned into pipes, which were described as larger at the outer end and smaller at the mouth end, much like their stone counterparts. No extant examples of Chumash wooden pipes are known, however. Only one instance of quasi-recreational smoking in a pipe was found in Harrington's notes. The Sun, a supernatural entity sometimes visualized as a man carrying a torch on his daily trip across the sky, was said to carry his pipe along with him. When he reached the *barranca* of noon, he sat down and took a smoke, then went onward refreshed.

For recreational purposes, eating tobacco was far more common than smoking it. Both of these practices were almost exclusively male activities. It was said that only men, youths, and a few "wild women" ate *pespibata*. Women never smoked, but they often kept a chunk of *pespibata* in their mouths, in front of the teeth behind the lower lip.

Tobacco was primarily smoked for ritual purposes by specialists using tubular steatite pipes. The smoke was used much like incense, as a purification or blessing, to treat various kinds of sickness, and to drive away rattlesnakes.

The stone pipe was regarded as a sacred object, and the owner took a vow never to carry it around just for fun or to use it in play. New pipes were sanctified with tobacco smoke by the *'alshuklash*, or pipe doctor, who acted as a priest. He inhaled from his own pipe and blew the smoke on the new pipe first. Then he blew smoke three times on the owner, starting at the feet and raising his head as he blew. He prayed during this ceremony to ask the mercy of *Kakunupmawa*, the Sun. For conducting any kind of blessing, the *'alshuklash* was paid with acorns, chia, bead money, or whatever the client had.

Another ceremony in which tobacco smoke was exhaled over ceremonial objects was described by Henshaw's consultant Justo in the 1880s. This was intended to bring rain, cause death to enemies, and result in other community benefits. Twenty pelican stones (carved, hooklike effigies) were arranged in a circle, five on each side, with a bowl of water in the center. Standing next to the bowl, the "medicine-man" smoked tobacco in a long, cigar-shaped stone pipe. He blew the smoke out, first over the water, then the pelican stones. The onlookers then moistened their faces with this water, which had been made holy by the ceremony (Henshaw 1955:152).

Chumash burial practices described by the explorer Pedro Fages involved alternately blowing tobacco smoke over the deceased person's body and singing (Fages 1937:33–34). A widow or widower might fear seeing a vision, which would be considered an evil omen. In that case the pipe doctor would come and blow smoke from a mixture of tobacco and *romerillo (Artemisia californica)* over his or her head and body. The person then lay down with a cross of *romerillo* branches under the head and could be assured of sleeping untroubled by visions. Both of these ceremonial uses of tobacco may be considered a form of blessing or purification.

Medicinal treatments were done by a team comprising the pipe doctor and two other old men. One had the task of finding out if anyone was sick, and the other sang if the sick person wished to hear any song. On three or sometimes nine consecutive nights, the three came to the house of the sick person. While the patient lay on a mat, the pipe was filled from the tobacco container and held vertically so the tobacco would not fall out. It was lit with a hot coal from the fire. The *'alshuklash* made passes over the patient's body first, and then breathed the smoke through rounded lips onto the person's face, temples, or chest.

After the treatment, the pipe was extinguished by putting a special wooden plug in the larger end. When the curers were done assisting the sick person, they emptied out the burned tobacco, cleaned the pipe, and inserted the wooden plug, which had been shaped to fit the opening. When not in use, the pipe was always kept plugged and wrapped in a small tule mat.

One famous pipe doctor, Iluminado, was known for taking three or four draws on the pipe without letting any smoke out between. He then came right up to the patient's face and let the smoke out all at once. It was said that a sick person—even one at the point of death—would sometimes actually recover as a result of that treatment. Such stunts were likely to cause ill effects in the pipe doctor himself. As a remedy, Iluminado crushed the dried flower of *rosilla (Helenium puberulum)*, mixed it with water, and drank it.

Tobacco was used medicinally in a number of other ways. A piece of *pespibata* could be chewed or mixed with a little water and applied to the skin as a topical anesthetic when piercing a child's earlobes, or rubbed on as a liniment for sore muscles. The *pespibata* pounded up in water was drunk to relieve pains in the stomach. According to the priest at Mission San Fernando in 1814, this mixture was fermented with urine before being taken for stomach pains or to heal wounds (Geiger and Meighan 1976:73). Presumably this remedy resulted in vomiting as already described for social tobacco drinking; it may have been effective in purging the body of internal parasites that caused abdominal pain.

Some consultants said that tobacco was taken to give a person courage and to increase his sense of "seeing things." Native tobacco species may have mildly hallucinogenic properties. An unusual, concentrated form of tobacco was prepared by boiling the whole plants in water to make a strong soup. After a while the plants were removed and the liquid further boiled down into a thick gum. This was allowed to cool in the cooking vessel and then shaped into sticks, which were pierced so that they could be worn on a cord around the neck. If a man just touched the tobacco-juice pendant to his tongue, it was believed to give him great bravery and strength, enough to climb mountains or even kill a bear.

Offerings made at shrines and other sacred spots were an important part of Chumash religious observances. Tobacco, bead money, and seeds such as acorns, chia (*Salvia columbariae*), *islay* (*Prunus ilicifolia*), and red maids (*Calandrinia*) were the most common offerings. In the Ventura region, for example, there was a special spring where the head of a family would collect water to use in curing a relative who was ill. A broken piece of *pespibata* was placed at each of the four cardinal directions, other offerings were strewn about the edge of the water-hole, and prayers were said before the water was taken (Hudson et al. 1977:65).

Alexander Taylor reported in the 1860s that the Santa Barbara Indians:

> [O]ften secretly built little temples of sticks and brush, on which they hung bits of rags, cloth, and other paraphernalia, depositing on the inside tobacco and other articles used by them as presents to the unseen spirits. This was the occasion of great wrath to the Padres, who never failed to chastise the idolators when detected (transcribed in Harrington 1986).

Tobacco was among the offerings directed at particular beings as well as places. When old women were gathering clams at Avila Beach, they would cast offerings of beads, feathers, and tobacco mixed with lime to the Swordfish passing offshore (Greenwood 1972:83). A Chumash legend relates that Coyote carried a ball of *pespibata* when he went to visit the Swordfish, who were old men living in a house under the ocean. When he got there, the door of the house opened its mouth just a little. Coyote gave it a bite of the *pespibata*, and the door opened its mouth wide enough to let Coyote inside.

There were a number of other beliefs about tobacco's capabilities. Teodoro Nawakmait told Fernando that if you want a good, brave dog, you should chew *pespibata* and put the saliva into the dog's mouth. If you do this, the dog will take care of you and die before he leaves you. And, on a number of occasions, José Juan Olivos told Harrington that the linguist would never be able to learn any old Indian custom unless he took *pespibata*.

Cactaceae | Cactus Family

Opuntia spp. | **PRICKLY-PEAR**

(B) khɨkhɨ' | (Sp) Nopal (plant, stems), Tuna (fruit)

(I) khɨ'

(O) tqɨ'ɨ

(V) khɨ'ɨl

In Chumash territory there are three native species of prickly-pear cactus: coastal prickly-pear (*Opuntia littoralis*), chaparral or tall prickly-pear (*O. oricola*), and Mohave tuna (*O. phaeacantha*). A fourth species, Indian fig (*O. ficus-indica*), was introduced from Mexico in mission times and has become widely naturalized. It is unclear whether one kind was preferred over another for fruit, edible pads, or useful thorns. The tendency of these species to hybridize with one another complicates the picture. The two local kinds of cholla cactus—another group of *Opuntia* species with cylindrical jointed stems—were not used by the Chumash.

To pick prickly-pear fruits, called *tunas*, Chumash people used a willow stick that had been split at one end. They cut the inside edges of the split, wedged it open into a Y, and tied it. They could then push this forked end against the cactus fruit and twist gently to dislodge it.

More bothersome than the conspicuous spines on this cactus are the tiny, sharp, hairlike bristles, called glochids, which must be removed before the fruits can be used. The Chumash took a sprig of *oreja de raton (Symphoricarpos mollis)*, coyote brush *(Baccharis pilularis)*, or some other plant and brushed the fruit while rolling it on the ground to remove these bristles. Often, especially on windless days, they would brush the bristles off the fruits of an entire plant one by one before picking them. If the glochids were not wet with dew, most of them would just fall off if the fruit was given a sharp blow. It was understood that one should never put cactus fruits into the basket or bag without first removing the bristles. Chumash myths portray Coyote gathering cactus fruit and being tricked into getting the spines in his eyes (Blackburn 1975:304–306).

The Chumash greatly relished *tunas* and usually ate them fresh. Most kinds they ate like figs, splitting them open with their fingernails from the base, then turning back the skin to expose the sweet, juicy pulp. In myths, people cut the *tunas* in half lengthwise with a knife. The variety that grew near *Simo'mo*, inland from Pt. Mugu, was eaten "straight," unpeeled. Left whole, the *tunas* were spread out on mats to dry in the sun. They were never boiled or roasted. Two other methods of preparation—straining the crushed fruit to make a beverage and grinding the dried seeds into flour—have been attributed to the Chumash (Weyrauch 1982:16) but were not mentioned by Harrington's consultants.

The red juice from prickly-pear fruits served as a paint pigment. For example, the musical bow, an instrument probably introduced to the Chumash, was painted red all over with fresh red *tuna* juice. Cactus-fruit juice was also mixed with pine pitch for painting the sun symbol on a whale vertebra to be displayed during the autumn gathering that preceded the winter solstice ceremony (Hudson et al. 1977:50–53). At Missions Santa Barbara and San Buenaventura, an Indian named Juan Pacífico painted pictures, wall murals, and ceilings with a paint made from red *tuna* fruit and egg white, cattle hoof glue, or pitch.

The flat stem segments of prickly-pear cactus, called *nopales*, exude a clear mucilaginous juice that was often used as a binder for paint pigments and in building construction during mission times (Webb 1952:107, 252). Large, mature cactus pads were baked until the thorns burned off, then chopped up and soaked in a barrel of water for five days. This liquid was used to help adhere adobe bricks to one another and to improve the water-repelling qualities of plaster and whitewash. The introduced "Indian fig" (*Opuntia ficus-indica*) was the type of prickly-pear preferred for this purpose, and this species was commonly planted as a living hedge in the colonial era. The sap from native *Opuntia* species could have served as a binder for mineral pigments in Chumash rock art, although the use of plant compounds in rock painting has not yet been established.

Fernando Librado said the Ventureño Chumash would dip the ball used to play the shinny game into this "cactus water." Perhaps it helped to consolidate the pounded *yerba santa* root (*Eriodictyon*) from which some balls were made. In more current medicinal use, the sap is believed to heal wounds and beautify the complexion; a tea of the flower is taken for the heart (Weyrauch 1982:16). Juanita Centeno said that one can split open the pads and bind them to any wound for fast healing without infection (personal communication 1978). This is similar to methods reported among the Paipai of Baja California and other tribes of New Mexico and Chihuahua (Ford 1975:247).

Very young cactus pads, or *nopales*, were eaten but seem not to have been a very important food for the Chumash. Harrington's consultants said that the Chumash did not eat pads of *O. ficus-indica*. The explorer Pedro Fages described the use of cactus pads as bait when fishing for sardines. The Chumash ground the cactus pads and threw them into large baskets in order to attract these fish, which they then caught in large numbers (Fages 1937:51).

Spines of prickly-pear cactus were used for piercing ears, after first massaging the earlobe between the thumb and forefinger. María Solares's grandmother plastered María's earlobes with chewed tobacco before piercing them, then treated the wound with an herbal wash. One might leave the cactus needle in the ear and tie it in place, or insert a small *carrizo* stick or sea urchin spine to keep the hole open. Both men and women had pierced ears, and some men also had their nasal septa pierced. Tattooing was done with cactus spines and pigment from charcoal or nightshade juice (see *Solanum*).

Historical accounts in 1602 and 1769 describe the Chumash fishing with cactus spines twisted into the shape of hooks, attached to lines, and baited with sardines (Vizcaíno 1959:16; Wagner 1929:242). Harrington's Chumash consultants denied that cactus-spine fishhooks were used (Harrington 1942:7), but perhaps this type of hook had gone out of use by their time.

Several of the same uses for *Opuntia* have been reported among southern California desert peoples. The Cahuilla formerly made basketry awls from cactus spines set in a piece of asphaltum for a handle (Barrows 1900:43).

Orthocarpus spp.: See *Castilleja*

Oxalidaceae — Wood-Sorrel Family

Oxalis albicans KUNTH subsp. *californica* (ABRAMS) EITEN — **CALIFORNIA WOOD-SORREL**

(B) 'aqnipshkáy — (no Spanish name known)

A note in Harrington's handwriting accompanied a specimen of this native plant, shown to Luisa Ygnacio in 1914. He noted that the name merely means "sour," and added, "The Inds. used to eat this much. Sour leaves. Ate raw."

Paeoniaceae | Peony Family

Paeonia californica TORREY & A. GRAY | **PEONY**

(B, I, V) mim | (Sp) Peonía

Harrington's consultant Simplicio Pico said peony was an ancient and important remedy used especially by women. When Lucrecia García had a baby, she was drinking a decoction of peony root as a hot tea before meals; a Mexican told her this remedy. Other authors say the root was used in the treatment of menstrual disorders and neuralgia; it could be chewed as well as made into tea (Bard 1894:6; Birabent n.d.). Spanish Californians took peony root, alone or in combination with other herbs, for indigestion or *empacho* (Blochman 1894c:163).

The Ohlone took a root decoction of *Paeonia brownii* to treat pneumonia and stomachache, and prepared the whole plants into a mixture drunk to cure indigestion or constipation (Bocek 1984:251). Various Indian peoples of southern California reportedly powdered the peony root and used it either in that form or as a decoction for colds, sore throats, and chest pain (Bean and Saubel 1972:98; Strike 1994:101). Peony has been used in Mexico to treat restlessness during fever (Ford 1975:269).

Viscaceae | Mistletoe Family

Phoradendron macrophyllum (ENGELM.) COCKERELL | **BIGLEAF MISTLETOE**

Phoradendron villosum (NUTT.) NUTT. | **OAK MISTLETOE**

(B) shlamulasha'w | (Sp) Muérdago

(I) stumuku'n

Several kinds of mistletoe grow in Chumash territory, but Harrington's consultants only mentioned those growing on sycamore and cottonwood (which would be *Phoradendron macrophyllum*) and on live oak (*P. villosum*). Fernando Librado said that women drank a tea of mistletoe to "make themselves sterile." A mission priest stated that mistletoe tea was drunk to bring on menstruation (Garriga 1978:24). This may in fact have been a method of inducing abortion, a subject which most Chumash consultants were understandably reluctant to discuss. Although the Catholic Church strongly discouraged abortion and contraception, both practices were reportedly common in mission times and probably earlier.

The Pomo took tea of bigleaf mistletoe to induce abortion and chewed the leaves to relieve toothache (Chesnut 1902:344). The Zuni drank a tea of juniper mistletoe (*P. juniperinum*) mixed with juniper bark as an aid to childbirth (Curtin 1965:40). The Cahuilla used the latter species to treat wounds and sore eyes, and made desert mistletoe (*P. californicum*) into a tea for unspecified medicinal use. They also ate the berries of both species, which, like the leaves, can be toxic (Bean and Saubel 1972:101).

Poaceae — Grass Family

Phragmites australis (CAV.) STEUDEL — **COMMON REED, CARRIZO GRASS**

(B) 'ekhpe'w — (Sp) Carrizo, Carrizo de Panocha

(I) 'ekhpew

(V) topo

The Spanish term "*carrizo*" was applied to three large grass species, two native and one introduced by Europeans (see also *Leymus condensatus* and *Arundo donax*). All three could be used interchangeably for certain purposes, such as house thatching, tubes for tobacco, cigarettes, knives, paintbrush handles, and game counter sticks. *Phragmites* was regarded as most suitable for certain kinds of arrows. It was also the only kind from which sugary honeydew was collected, hence the name *carrizo de panocha*, or "sugar *carrizo*."

Phragmites grows in a few spots in the Santa Ynez River watershed and in the Cuyama, Santa Clara, and Ventura River valleys. This plant may have been more widespread in the early days, for Fernando Librado said it was seen in Ventura "long ago." Cattle and horses have reportedly destroyed stands of *Phragmites* in Kern County (Twisselmann 1967:180), and the same may have occurred in Chumash territory. The Ventureño Chumash gathered the reeds at *Sitoptopo* ("place of much *carrizo*," now seen in the place name Topatopa) on San Antonio Creek, near Ojai. They tied the canelike stems into bundles to carry them home.

The *carrizo* stems had to dry completely with the bark left on before they could be worked into arrows. If the stems were straightened while still green, they would become crooked again in drying. The arrows were about two and a half feet long. The arrow maker inserted a hardwood foreshaft of toyon (*Heteromeles arbutifolia*) or, according to

Simplicio Pico, sometimes of *romerillo (Artemisia californica)* into the larger, basal end of the stalk. The basal end was stronger, and this way the leaf joints would be pointing toward the back of the arrow. Three feathers were attached at the opposite end, or nock. One could hunt rabbits, birds, and other small game with an arrow that just had the plain wooden foreshaft sharpened to a point, or one could attach a chert point to the tip for hunting deer or for war arrows. Particularly in war arrows, the joints of the *carrizo* stems were not polished but instead left rough so that the arrows would be more difficult to extract.

A few other minor uses of *Phragmites* stems were described by Harrington's consultants. After a broken arm had been set, it was immobilized in a splint made from *Phragmites* stems. To make the splint, *carrizo* pieces ten inches long and one-quarter inch in diameter were twined together into a sort of mat, which was wrapped twice around the arm and tied in place. Fernando Librado saw these splints being used by Spanish Californians.

Dancers in the Devil Dance and the Barracuda Dance held small *carrizo* whistles, two to three inches long, in their mouths. People who went seed gathering in the mountains often took whistles along so they could give warning if a bear was seen. These whistles may also have been made from *Phragmites.*

A knee-length apron worn under the belt from which game was carried was said to be made from *carrizo* or *Juncus.* The material may have been shredded for this purpose but was not twisted into cordage. Harrington disagreed with Barrows's (1900:47) assertion that cordage was made from *Phragmites.*

Phragmites is perhaps best known as the source of a sugary substance that was a favorite delicacy of the Chumash and various other Indian peoples. Harrington's consultants discussed the preparation of this sugar, called *mow.* They all mentioned acquiring the finished product from the Yokuts, particularly at festivals. María Solares, who had both Chumash and Yokuts relatives, had seen the process of making this sugar. She said the plants ripened in June, at which time people cut the whole plant above the root, dried it, then cleaned it and pounded it up. They made the resulting mass into cakes, wrapped them in tule (*Scirpus*), and brought them from the Tulare country to Santa Ynez for sale.

According to Fernando Librado, they used only the young shoots. The right time for gathering them was variable, he said; some Indians were guided by the phases of the moon. They mashed the shoots to a fine pulp and allowed this to sit for a period of time, but Fernando did not make it clear whether the shoots were dried before pounding or afterward. After a while it turned into *panocha*, or sugar, which was then formed into lumps or balls the color of dirt or lead. Some people dried it in the shade to make the balls a more bluish color. Fernando said they carried the lumps of *mow* in a special tule container about a foot long, woven at one end and tied at the other, and with an oval cover attached by hinges.

The Chumash cut pieces of this *carrizo* sugar off the bigger lump and ate it, a bite at a time, between mouthfuls of chia (*Salvia columbariae*). Or, because the *panocha* tended to stick to the teeth, one might take a drink of chia mixed in water between bites. "Oh, how good it was!" consultants said. "It is the nicest thing you ever tasted." The mission priests gave permission for Chumash people to go out gathering *carrizo* for sugar, because they knew the people were so fond of it.

Information collected from other California Indian groups indicates that the sugar from *Phragmites* is actually a honeydew deposited by aphids on the leaves and stalks. Chumash consultants do not seem to have understood this, perhaps a further indication that they most often acquired *carrizo* sugar from other groups rather than preparing it themselves.

Among the Tubatulabal, people cut the plants in July or August, spread them out in the hot sun to dry, and beat them with sticks to dislodge the sweet granules. These they mixed with cold water to make a stiff dough which they spread on a tule tray, then put away to dry for six or seven days. They knocked small lumps of the sweet off the hard, brown lump with a rock and ate them dry with chia gruel (Voegelin 1938:19). The Yokuts followed a similar harvesting method but were seen to eat the sugar in its granulated form rather than in molded lumps (Latta 1977:443–445).

Zosteraceae — Eel-Grass Family

Phyllospadix torreyi S. WATSON

SURF-GRASS, SEAGRASS

(B) shkash
(V) chkapsh
(Sp) Zacate del Mar

This marine flowering plant was described by Harrington's consultants as "green hair seaweed," and they mentioned only a few ways in which it was used. The Chumash would gather it, when available, to eat in the sweathouse: if a person was overcome by the heat in the sweathouse and began to feel faint, he took a fistful of this "hair seaweed" and put it in his mouth. When fish were being dried, surf-grass, freed of sand, was draped over them for the first few nights. People also cut the plant and bundled it up into knee pads used when paddling plank canoes. Luisa Ygnacio said that non-Indians sometimes used surf-grass as a mattress, but she had never seen it woven into a mat or anything like that.

Archaeological evidence shows that Chumash use of *Phyllospadix* was much more extensive than Harrington's consultants indicated. This plant was very important to the island Chumash as a substitute for a number of common mainland species that were not available on the Channel Islands (Timbrook 1993). For example, the remains of houses thatched with surf-grass have been noted in a number of island sites (Rogers 1929:315, 332; Orr 1968:212, 217, 219). Some of this material was said to be a related species, eel-grass (*Zostera marina*), but its leaves are usually much wider. Samples of remarkably well-preserved *Phyllospadix* thatch from archaeological house remains on Santa Rosa Island have been radiocarbon dated to 1860 ± 340 years before present (Orr 1968:212). Resistant to decay and to fire, surf-grass was an excellent substitute for tule (*Scirpus*) and cattail (*Typha*), which are relatively scarce on the islands.

The remains of what appear to be surf-grass skirts are preserved in museum collections from the islands. The strands are laid parallel and fastened together with several rows of twining at the top, leaving the ends hanging loose. Skirt weights—small, clove-shaped globs of asphaltum, each containing the tip of a single blade of grass bent sharply to one side—have also been found in large concentrations (Rogers 1929:407–408). These weights were presumably molded around the ends of the surf-grass at the hemline to help this relatively light material to hang more modestly and move more gracefully. Though skirts of shredded plant material (tule, cottonwood bark, willow bark) were described by Harrington's consultants, surf-grass skirts were not used by the mainland Chumash.

Many other types of objects made from *Phyllospadix* have been noted in archaeological investigations on the islands. These include fragments of twined mats, plied and braided cordage, nets, and fishing lines to which fishhooks were secured with asphaltum (Hoover 1971:166–167; Hoover 1973a:6–7). Surf-grass was apparently a common island substitute for Indian-hemp (*Apocynum*) in the making of cordage for all sorts of uses. Closely twined baglike containers may have been used instead of certain kinds of *Juncus* baskets, for remains of basketry are scarce on the islands (Timbrook 1993:53, 55).

Common as surf-grass artifacts are in island sites, they are very rare in mainland archaeological deposits, even though the plant is common around rocky areas in surf along both mainland and island coastlines (Smith 1998:127). Mainland Chumash peoples must have had easier access to other plant species that they regarded as more suitable for most purposes.

Pinaceae — Pine Family

Pinus spp. (except *P. monophylla*) — **PINE** (general)

(B, I) tak, tomol — (Sp) Pino

(V) tsɨkɨnɨn

With the exception of pinyon pine (*Pinus monophylla*), the Chumash did not have a separate name for each species of pine recognized by modern-day botanists. Similarly, Shoshonean peoples commonly divide pines into two categories: "*piñón*" and "pines other than *piñón*" (Zigmond 1941:37). The Barbareño and Ineseño Chumash further divided non-pinyon pines into two main types, *tomol* and *tak*. Jeffrey pine (*P. jeffreyi*) and ponderosa pine (*P. ponderosa*) comprised the *tomol* group. The *tak* category is less clear, but it seems to have included Coulter (*P. coulteri*), limber (*P. flexilis*), sugar (*P. lambertiana*), gray (*P. sabiniana*), Torrey (*P. torreyana*), and bishop pine (*P. muricata*) (Hudson, Timbrook, and Rempe 1978:48). All of these species grow in Chumash territory, though some are restricted to high interior mountains.

In Spanish, just as in popular English usage, the word for "pine" may refer to more than one genus of coniferous trees. For example, Harrington's consultants used the term "*pino colorado*" to refer to both Jeffrey pine and coast redwood (*Sequoia sempervirens*).

Though *Sequoia* was generally the preferred species for making plank canoes, Jeffrey pine is thought to have been one of several other woods used (Hudson, Timbrook, and Rempe 1978:46–50). Juan Justo recounted a myth in which Coyote went to a place called *Tomto'mo'* to buy *tomol* pine boards for making canoes. There was an old man there who had the trees all cut and ready to sell (Blackburn 1975:207–211). Both Fernando Librado and Henshaw's consultant Justo speculated that pines on Santa Cruz Island (bishop pine, *Pinus muricata*) might have been used in canoe construction (Henshaw 1955:151).

Several years ago, a pine-wood stirring paddle which had apparently been employed in the preparation of mush from wild cherry pits (*Prunus ilicifolia*) was found in a cave cache on Santa Cruz Island. This artifact was made from sugar pine (*P. lambertiana*), which could have been obtained as driftwood or through trade with the interior mainland (Timbrook 1980).

Pine logs were brought from the mountains for use as roof beams in the construction of Mission San Buenaventura. Fernando Librado said that "white pine," the kind found in Santa Paula Canyon, was full of pitch and hard to plane into flat boards, and it also rots when used in construction.

Many northern California peoples ate the large nuts of gray pine and sugar pine (Farris 1982). The Chumash may have eaten the nuts of these and perhaps other species of pine, as Luisa Ygnacio indicated that *tak* pine had seeds that were good to eat.

Some contemporary Chumash people throw pine needles into hot bath water as a treatment for rheumatism (Weyrauch 1982:15). Modern Chumash weavers have also adopted the use of pine needles, especially those from Torrey pines, as a foundation material in open coiled basketry.

For uses of pine pitch, see *Pinus monophylla*.

Pinaceae — Pine Family

Pinus monophylla TORREY & FRÉMONT — **ONE-LEAF PINYON**

(B, I, V) posh — (Sp) Piñón

The Chumash used the wood and nuts (technically, seeds) of various species of pines but singled out *Pinus monophylla* as being the best source of pine nuts for food and wood for making bows. Pine pitch, an important adhesive and caulking material, was probably also obtained primarily from this species. The Spanish term "*piñón*" refers to both the tree and the nuts.

Rather than waiting for the seeds to ripen fully and fall from the cones in late September or October, the Chumash most often collected them in August. Every year many people from coastal villages traveled inland to San Emigdio, Mount Pinos, *Mat'ilha*, Cuyama, and Tejon to gather pinyon nuts. There were numerous shrines in the mountains along the pinyon gathering journey, and people always made sure to leave something as an offering when they passed by. Taboos prohibited harvesting the nuts in certain sacred places. These expeditions continued well into the mission period, and the priests gave the Indians permission to go.

The Milky Way was known as "the way of the *piñón* gatherers," for it was white like the insides of pinyon nuts (Hudson et al. 1977:35). It was said that when the Milky Way led to the north, there would be many pinyon nuts. Perhaps this referred to their availability only at a certain season. This conspicuous band of stars does trend toward the northern horizon in the evening hours during September and October. The quantity of nuts produced by pinyon trees depends on many factors, however, and varies greatly between locations and from one year to the next. The Chumash believed that in a dry season pinyon nuts had "too much heat" and could make one who ate them sick.

At the *piñón* gathering camp, men climbed the trees early in the morning, beat the branches with short poles to knock the cones loose, picked the ones they could reach, and threw them down to the women below. Later in the day, people stayed on the ground and employed longer poles to knock the cones from the trees. If they climbed then, when it was warmer, they would get covered with sticky pitch.

Women collected the cones in basin-shaped baskets and piled them up to be roasted. This was done to eliminate the pitch and to facilitate removing the seeds from the cones. The men built an above-ground roasting oven by digging four post holes in a square and erecting sticks at the corners. They dug out the bottom slightly and filled the entire inside of this bin with pinyon cones. Sometimes they merely placed the cones on a bed of pinyon branches, without building any kind of pit or bin. Then they placed dry pinyon sticks around the sides and top, and set the whole pile on fire. When it had burned for a while, they covered the cones with earth and ashes.

The resin burned and the cones would open. Each person would take one cone at a time from the pile and strike it against a rock or press it between the hands, twisting and shaking, which made the seeds fall out. They continued opening the cones in this way until all were done. Cones should never be left buried in the ashes all night; if they could not all be opened in the afternoon, they must be taken out of the ashes and carried over to the campsite to be opened that same evening. The seeds were set out to dry for two or three days, then winnowed to remove any empty shells. The people carried the pinyon nuts home in baskets or in buckskin bags, two or three such bundles in one carrying net.

The Chumash generally cooked pinyon nuts and did not eat them raw. They toasted the nuts in baskets with hot coals, cracked them open with mortars and pestles or on flat rocks, and winnowed them to remove the shells. They might then further pound the nutmeats into a fine, dry flour or toast them again and eat them whole.

Studies of the nutritional value of pine nuts have found that 72 percent of the seed of *Pinus monophylla* is edible material. The hulled kernels contain 53.8 percent carbohydrates, 23 percent fat, and 9.5 percent protein. At 2,213 calories per pound, they are a good source of energy (Farris 1980:134). Harrington's consultants noted that Chumash pine nuts were larger and richer in flavor than the Mexican ones.

The Chumash made "baskets" of strung pinyon nuts. For this they used tree-ripened nuts that had fallen and were picked up off the ground, rather than the nuts they had roasted in their cones. They parboiled these ripe nuts to soften them, pierced them with a sharpened wooden needle, and strung a few at a time. After three or four nuts had been put on the string, a loop or stitch was taken to fasten them to the previous part. Gradually, this grew into a container the size and shape of a globular trinket basket with a narrow mouth and a handle at the top. More strings of nuts were hung from the sides, and the inside was filled with loose pine nuts. These *piñón* baskets could be used right away or kept for a long time. The Kitanemuk as well as the Chumash made them (Hudson and Blackburn 1983:71–72).

Chumash people usually strung pinyon nuts for both storage and trade, and they kept strings of nuts in the house to be eaten any time. A person would crack one nut, pull it off the string and eat it, then crack and eat another, and so on. For trade, the Chumash measured strung *piñón* nuts around their hands in the same fashion as shell-bead money. These nuts were quite valuable: one strand of *piñones* was worth three of shell-bead money.

On festive occasions, strings of pinyon nuts could be seen in almost all Chumash houses. At fiestas, women would take bundles of strung *piñones*, break the strings, and toss them to the crowd. People would pick them up and eat them on the spot. This was done particularly in the mourning ceremony, which was held every two or three years. It was timed to occur after the *piñones* had been harvested, usually in September. At the mourning ceremony procession, the participants scattered pinyon nuts, whole acorns, chia seeds, and shell-bead money. Pinyon nuts were also given as offerings at the winter solstice ceremony and at shrines.

To make an ordinary bow, Chumash people merely stuck a live-oak (*Quercus agrifolia*) shoot into the fire and bent it. Juniper could be used, but it had too much pitch and soon rotted. For the very finest bows, said Fernando Librado, they used only pinyon wood. These fine bows were about three feet long and had a strong recurve, a form which the bow maker achieved by bending the wood after treating it with hot water. He glued the sinew from the back or legs of a deer onto the back of the bow with pine pitch.

Pine pitch was of great importance to the Chumash as an adhesive and sealant. They observed that, particularly in summer, certain trees exude considerable pitch where a branch has broken off. Ventura people collected quantities of it from the *Mat'ilha* area, for example, when they went to gather *piñón* nuts. Some pitch was hard, in lumps, and some was soft like dough. They packed it in sacks woven of *tok (Apocynum)* fiber, carried it home, and kept it until needed in canoemaking or for gluing anything. In the mission carpenter's shop, containers were always kept ready to collect pitch that ran out of the wood they worked with.

One of the principal uses of pine pitch as a caulking compound was in the process of building plank canoes, described by Fernando Librado (see Hudson, Timbrook, and Rempe 1978:50–53, 78–80). To form the hull, the canoe makers glued the planks together one row at a time, using a mixture of pitch and asphaltum (a form of tar). To prepare this mixture, called *yop*, they half filled a large soapstone *olla* with tar and placed it on the fire. Then they added two double handfuls of pulverized pitch to the melted tar and stirred briskly. The correct proportions were important—insufficient pitch would make the tar brittle, but too much would prevent it from drying hard. When they had glued the planks in place and left them to dry for three days, they drilled holes on each side of the seams and lashed the planks together with string made from Indian-hemp (*Apocynum*) fiber. They waxed this cordage with a mixture that was similar to the adhesive compound but with a higher proportion of pine pitch to retain flexibility. When the whole canoe had been assembled, they caulked all the seams again by packing them with inner pith from tule (*Scirpus*) stalks and coating them with the *yop*. Finally, they sealed the canoe with a paint made from tar, red ochre, and pine pitch.

The Chumash also used a mixture of tar and pine pitch to attach arrow points to hardwood arrow foreshafts. They usually caulked basketry water bottles with tar, but pitch may have been added to the pulverized asphaltum. The Chumash employed either material to coat drinking cups, which were small basketry bowls woven of *Juncus;* the pitch-coated ones were red on the inside and smelled of pine (Heizer 1970:64, 67).

Pine pitch sometimes served as a binder for paint. During the sun ceremony held at the fall harvest festival, a sun symbol was painted on a whale vertebra disk with purple juice of cactus fruit mixed with pine pitch (Hudson et al. 1977:50).

To make a jet-black pigment for face painting, dry pinyon twigs were burned inside a small, boxlike structure of stones. The soot that accumulated on the roof stone was scraped off from time to time. When enough had been collected, it was kneaded with deer-bone marrow to a doughy consistency. This paint was especially prized because it did not rub off.

A poultice of pine sap was applied externally for colds and pains (Birabent n.d.). Some people also chewed pine pitch like gum.

Boraginaceae — Borage Family

Plagiobothrys nothofulvus (A. GRAY) A. GRAY — **POPCORN FLOWER**

(B) k'á'nay — (no Spanish name known)

Luisa Ygnacio described this plant as eight inches tall, somewhat prickly, with very small white flowers. She saw some growing in San Roque Canyon, behind Santa Barbara, and remembered that she and other children used to rub the root on their faces and hands to color them red for fun. Other California peoples did the same (Mead 2003:314–315).

Plantaginaceae — Plantain Family

Plantago spp. (Introduced) — **PLANTAIN**

(no Chumash name known) — (Sp) Lantén

Several species of plantain are found locally, but the two most common are broad-leafed weeds introduced from Europe, *Plantago lanceolata* and *P. major*. Chumash people noted that plantain was similar in appearance to *yerba del manso (Anemopsis)*—a description that fits both of the nonnative *Plantago* species—and used it medicinally. They applied the leaves to sores to draw out the poison, or dipped the leaves in hot olive oil and placed them on the cheek to treat toothache. A sick person might also place the leaves in water for bathing. Other sources mention *lantén* as a poultice for boils, sore throat, abscesses, and swellings, most often after being heated in olive oil or lard (Birabent n.d.; Gardner 1965:299; Garriga 1978:17, 40).

Although the native California plantain (*P. erecta*) was used medicinally by some California peoples (Strike 1994:113), it is likely that the medicinal use of *Plantago* was not indigenous to the Chumash. *Lantén (Plantago major)* leaves are crushed and applied externally for cuts, bruises, wounds, and headache by several groups in Mexico, Baja California, and New Mexico (Ford 1975:220–221).

Platanaceae — Sycamore Family

Platanus racemosa NUTT. — **WESTERN SYCAMORE**

(B, V) khsho' — (Sp) Aliso

(Cr) qsho'

(I) shonush

(O) teksu

(P) 'aqsho'

In California, the Spanish word "*aliso*" most often refers to sycamore. When translating early historic documents, some writers have been misled by the fact that many Spanish dictionaries translate *aliso* as "alder," an entirely different tree (*Alnus* spp.). Alisal Canyon in Santa Barbara County was named for its numerous large sycamore trees. It is curious that Harrington recorded a name for this tree in the island Chumash language. Sycamores are not native to the islands, but some were planted on Santa Cruz in about 1937 (Junak et al. 1995:229–230).

The Chumash made wooden bowls from round burl-like growths found on the trunks of some sycamore trees. They also used the gnarled branches, which naturally grow in sharp angles or, as the Chumash called them, elbows. They removed the appropriate portion from the tree and worked it with stone tools right away while it was still green; if they allowed it to dry, even for two or three days, it would crack (Hudson 1977:16). They hollowed out the inside of the bowl first, then turned it over and shaped the outside. When the bowl was finished, they smeared it all over with a paste of red ochre mixed with squirrel fat and set it in the sun. They used dry stems of scouring rush (*Equisetum*) to polish the bowl, then applied more oil, and repeated this process several times.

These wooden bowls were very smooth and round, and the wood grain was mottled in appearance. The Chumash made them in various sizes for different purposes, from chia containers less than a foot in diameter to large ones the size of washbasins. The early Spanish explorers noted that Chumash wooden bowls and trays were so finely crafted as to appear to have been turned on a lathe (Crespí 1927:159; Fages 1937:35).

In the Barbareño and Ventureño languages, the name for sycamore tree, *qsho'*, is also the word for wooden bowl. Other woods were sometimes used to make bowls, including willow, toyon, cottonwood, and alder. Examples of Chumash oak bowls have been preserved at the Musée du quai Branly (formerly Musée de l'Homme) in Paris, and an

WESTERN SYCAMORE
Platanus racemosa

example of a bay-wood bowl can be seen at the Santa Barbara Museum of Natural History.

Poles of sycamore were sometimes used like cottonwood or willow in house construction (see *Populus*, *Salix*).

Harrington's Chumash consultants also described wagon wheels made from cross-sections of sycamore trunks during the historic period. These solid wheels were probably the type used on the two-wheeled oxcart, or *carreta*, which was very important in transportation of goods in Hispanic California. A tree with a round trunk of the desired size was chopped down, and sections about eight inches thick were sawed from it with two-man saws. A hole was cut in the center of each disk. To keep them from cracking, the wheels were soaked in waterholes in a creek for a month or two. This was done in the summertime so that floods would not wash them away. Then the wheels were mounted onto the *carreta*'s oak axle with pegs.

The Ohlone wrapped bread with sycamore leaves during baking. They also ate the inner bark, and made a medicinal tea from an unspecified part of the plant (Bocek 1984:249). The Diegueño treated asthma with a decoction of sycamore bark (Hedges and Beresford 1986:30).

Salicaceae — Willow Family

Populus fremontii S. WATSON — **FREMONT COTTONWOOD**

P. balsamifera L. subsp. *trichocarpa* (TORREY & A. GRAY) BRAYSHAW — **BLACK COTTONWOOD**

(B) qweleqwe'l — (Sp) Alamo
(I) qweleqwel
(V) khwelekhwel

Spanish dictionaries translate *álamo* as "poplar," an English term for members of the genus *Populus*. In California, these trees are more often called "cottonwood."

The Chumash made their house poles of cottonwood, sycamore, or willow. They selected the tallest possible poles and cut them, rather than burning them down. They removed the leaves and twigs and charred the base of each pole to reduce rotting. To build the house, they dug holes in a circle, leaving a wide space for the door. They planted the bases of the house poles in these holes, then drew the tops together and tied them. After this, they lashed on smaller poles horizontally and then attached the thatching (see Hudson and Blackburn 1983:323–337). Eventually the buried part of the pole did begin to decay, but it was not necessary to replace the entire pole. Next to the old rotten base, they would just set a new post into the ground and lash it to the old pole.

Renowned as skilled woodworkers, Chumash people made some of their wooden bowls, trays, and containers of cottonwood. For example, the sunstick used in the winter solstice ceremony was kept in a box of cottonwood lined with feather down and inlaid with shell (Hudson et al. 1977:57). They also sometimes made dugout canoes from cottonwood trees, according to Fernando Librado, but it was difficult to find logs of sufficient size. Dugouts were not common in the Chumash region, at least in historic times, and were used in estuaries rather than on the open ocean (Hudson, Timbrook, and Rempe 1978:31–36).

In Chumash oral tradition, Old Man Sun carried a firebrand as he journeyed daily across the sky. This torch was made from the inner bark of a tree much like a cottonwood, which grew only in the sky. The inner bark fiber was rolled tightly so that it would burn slowly on the Sun's journey (Blackburn 1975:92). Perhaps the Chumash themselves used similar torches of cottonwood bark.

Poor women wore skirts made from cottonwood fiber, as opposed to the tanned animal hides, including deer, fox, and sea otter, that wealthier women wore. To make a skirt from cottonwood, they stripped the bark from the green tree, dried it, and then softened it by bending and rubbing it with their hands. This fiber skirt was worn belted at the waist and consisted of two parts, a front flap and a back flap. Both these parts were woven across the top with cordage of the same cottonwood fiber, leaving the remainder to hang in strips like long fringe that reached the knees. Skirts like this could also be made of tule (*Scirpus*) or the inner bark of "curly willow" (*Salix laevigata*).

To stabilize and cushion objects they carried on their heads, the Chumash fashioned cottonwood bark into a ring. They never placed storage baskets directly on the ground, but put a layer of cottonwood bark, other bark, or poles under them so that rodents would not gnaw through the baskets and get the seeds they contained.

Simplicio Pico, who had made green cottonwood bark into a hot medicinal tea to bathe broken or bruised limbs, told Harrington that it was very effective in providing relief. Many other Indian peoples, from northern California to Baja California, also used Fremont cottonwood leaves or bark as a poultice or wash for bruises and cuts (Chesnut 1902:330; Ford 1975:120).

Potentilla glandulosa: See *Horkelia cuneata*

Rosaceae | Rose Family

Prunus ilicifolia (NUTT.) WALP. | **HOLLY-LEAVED CHERRY**

(B, I) 'akhtayukhash | (Sp) Islay

(Cr) wam

(O) chto

(V) 'akhtatapɨsh

The Spanish word "*islay*" (pronounced to rhyme with "fly") is derived from *slay'*, the Salinan Indian name for this plant (Harrington 1944:38). Like the Chumash terms, it refers to both the plant and its cherrylike fruit. *Prunus ilicifolia* is the most common wild cherry in coastal California south of San Francisco Bay, and it was used as food by every group in whose territory it occurs (Timbrook 1982:163). Like most other Indian peoples, the Chumash did eat the fruit pulp but made far greater use of the inner kernel or seed, which was a highly valued food source. As several consultants said, "the kernel was the really esteemed part of the *islay*."

Islay fruits were picked in late summer. One must never beat the tree with sticks or climb up in it, the Chumash said, for it is very delicate. They pulled the ripe, red fruits one by one from the trees, and also picked up those that had already fallen to the ground. They picked the green fruits, too, but kept them in a pile separate from the others.

For gathering *islay* fruit, the Chumash made special network bags from cordage of Indian-hemp (*Apocynum*) or nettle (*Urtica*). They wove the meshes just finely enough to keep the *islay* from falling through and attached a hoop of willow or sumac wood at the mouth to hold the bag open. *Islay* bags varied in size; the ones men used were large, about eighteen inches long and a foot in diameter when full. As a person gathered *islay*, the bag hung on the chest, suspended from a woven carrying strap that passed behind the neck or over the shoulder. When the bag was filled, it could be carried on the back with the strap as a tumpline over the head. The fruit was emptied from the gathering bag into a larger, closely woven sack which had drawstrings at both ends. The *islay* could be transported in this sack, but people would often clean it near the gathering site and bring home only the pits.

They piled up the fruit on a clean-swept earth floor or in a basin-shaped basket and left it for several days until the pulp rotted. This they rubbed off with the hands or washed off in the creek, leaving just the pits. Next, they heated stones in the fire and placed them in a basket of water. The cleaned *islay* pits could be simply immersed in the hot water, or the water could be poured over them, or they might be boiled for a short time. After this preliminary treatment, the pits were allowed to sit overnight or spread out to dry in the sun for two or three days. Then the pits were cracked open with a rock and the kernels were removed. Some of Harrington's consultants said that at this point the shells from the green *islay* fruit, which had been kept separate, were burned and their ashes mixed with water to a doughy consistency, then molded into cakes for use later in the cooking process. Only the shells from green *islay* were used for this purpose.

After this, it was all right to mix the kernels of both red and green *islay* fruit together. These dry kernels could be stored indefinitely and were usually kept in the home in large baskets. The storage baskets for *islay* and other seeds were two and a half feet wide and two feet high. People made sure to place them on sticks or on tule mats, not right on the ground.

Before one could eat the *islay* seeds, they had to be leached to remove the hydrocyanic acid which makes them not only extremely bitter but poisonous. If *islay* was not properly prepared, the Chumash said, it could make a person very sick. According to one of Harrington's consultants, in historic times when the Indians got tuberculosis and were spitting up blood, they blamed it on the *islay*. The toxicity of *Prunus* pits is discussed in more detail elsewhere (Timbrook 1982).

There was some variation in the methods of *islay* preparation, leaching, and cooking described by Harrington's consultants. Three processes were reported for leaching *islay*, and they all differ somewhat from the way acorns were leached (see *Quercus*). One might place the whole kernels in a sack and repeatedly dip it into hot water (Saunders 1920: 57–59). Or the *islay* could be mashed and placed in a basket in the creek to let water run through it. The third method was to place the islay kernels in a steatite *olla*, cover them with cold or lukewarm water and heat them on the fire. When the water had nearly boiled, it was poured off, replaced with clean water, and heated again. With relatively fresh kernels one should change the water three times, but only twice with old ones because they were less bitter.

After completing the leaching process, the Chumash would boil the *islay* kernels in water until done. Since cooking took several hours, they did this in the steatite *olla* over direct heat rather than in baskets with hot stones. As the kernels were boiling, some cooks took the ash cake they had previously prepared from the burned green *islay* hulls and added an amount equal to the last joint of two fingers. They believed this would counteract any residual bitterness.

When the water had nearly boiled away and the *islay* had become soft, they stirred and mashed it with a wooden paddle. Using the paddle or an abalone shell, they scooped the mashed *islay* a little at a time out of the pot and molded it between the hands into round balls, from the size of a biscuit up to five inches in diameter. These were rolled in pinole flour of juniper or grass seeds and placed on a basket tray.

The prepared *islay*, called *shukuyash*, resembled beans in both flavor and reddish color. The Chumash consultants all agreed that it was a good-tasting and prized food. Its nutritional role in the diet is not yet understood. Analysis of raw *Prunus ilicifolia* seeds has yielded contradictory results: one study determined that the kernels contained almost no starch, yet another found them to be over 70 percent carbohydrates (Earle and Jones 1962; Gilliland 1985:46). The composition of the seed kernels prepared the way they would have been eaten traditionally has not been studied.

Islay balls could be kept as long as a week and were commonly offered to visitors. The Chumash often ate them with roasted squirrel, gopher, or other meat. Coyote had them for breakfast in one Chumash myth (Blackburn 1975:216), and they are mentioned in the song of the Santa Rosa Island Fox Dance (Hudson et al. 1977:71). Offerings of *islay* seeds or prepared *islay* balls were an important part of most ceremonial occasions (Hudson et al. 1977; Hudson, Timbrook, and Rempe 1978: 141).

The Spanish explorers described a number of foods eaten by the Chumash and other California Indians, and at least two of these early accounts may refer to *islay*. The diarist of the Cabrillo voyage referred to a white seed the size of maize, from which good "tamales" were made (Bolton 1925:30). This sounds very much like the *islay* balls of the Chumash. With their shells still on, holly-leaved cherry seeds are white and about the size of maize kernels prepared into hominy, from which Mexican tamales are made. Longinos Martínez mentioned a large seed called *silao* which, despite its bitterness, was washed, dried, and roasted to make one of the Indians' most important foods (Longinos Martínez 1961:46). This name closely resembles the Salinan name *slay'*. On the other hand, the even larger seed of buckeye (*Aesculus*) was prepared in a similar way by peoples to the north of the Chumash.

Islay was a significant trade item between Chumash groups of different regions. Some people said that the islanders did not bother to gather their own, but instead made shell-bead money to sell to the mainland Chumash in return for as much *islay*, acorns, chia, and *pil (Calandrinia)* seed as they could carry. A larger-fruited, less spiny-leaved form of holly-leaved cherry (subspecies *lyonii*) is common on the larger Channel Islands. The island inhabitants surely harvested some local fruit, but their reliance upon mainland sources indicates *islay*'s widespread importance as a food resource (Timbrook 1993:52). In trade, the seeds were measured with women's basketry hats; one hatful of *islay* was worth two of acorns.

Harrington's consultants remembered getting permission to gather *islay* on Alisal Ranch in the Santa Ynez Valley. Different families went at different times to gather their own supplies. They met at night, built fires to keep the bears away, and had singing, dancing, and praying at the camps. Another common gathering location was in Mission Canyon. The Chumash name for a place near Mission Santa Barbara was *'akhtayukhash, islay* (Applegate 1975b:25). *Islay* pits have not been used as food in recent times (Gardner 1965:285–286).

Occasionally the Chumash did eat the *islay* fruit pulp, either fresh from the tree or mashed and spread out on a flat rock in the sun. When the pulp had dried, they rolled it up and ate it like fruit leather. Modern Chumash remember the fruit being fermented to make a drink called "tweeswin" (Weyrauch 1982:20). Somewhat like a fruit wine, *tesgüino* was a mission-era introduction from Mexico, as the Chumash made no fermented beverages in prehistoric times (Geiger and Meighan 1976:89).

Polypodiaceae — Fern Family

Pteridium aquilinum (L.) KUHN. — **BRACKEN FERN**

(B) qɨch — (Sp) Palma, Manita

(V) kich

The Chumash applied the Spanish common names *palma*, *palmilla*, and *palmita* to several species of ferns, not just bracken. Elsewhere in California, these names can refer to palm trees or to *Yucca* species.

In some areas, particularly inland, the Chumash used bracken fern for house thatching. They attached the fronds to the horizontal crosspieces of the house framework in the same manner as they did *carrizo*, tule, or cattail. The tip ends pointed downward to shed water.

Bracken fronds played a role in roasting food in earth ovens. For example, yucca hearts would be put on hot coals and stones in a pit, covered with a layer of bracken fern fronds and a layer of dirt, then left to cook overnight. The fern fronds protected the food from dirt. If several people were cooking yucca hearts or stalks in the oven at the same time, they separated each person's portion from the others with a layer of bracken.

In rare instances, Chumash coiled baskets may contain sewing strands of bracken-fern rhizome, dyed black by burial in mud (Dawson and Deetz 1965:202). This material is much more commonly seen in baskets made by other groups to the north and east of the Chumash, including Yokuts, Mono, Pomo, Washo, and Paiute.

The Cahuilla and some other peoples ate the young shoots of bracken (Bean and Saubel 1972:121; Heizer and Elsasser 1980:250), but apparently the Chumash did not do so. Fern "fiddleheads" were a more important native food in the Pacific Northwest than in California.

Fagaceae — Oak Family

Quercus spp. — **OAKS** (general)

(no generic Chumash name known) — (Sp) Encino, Roble

The Spanish term "*encino*" refers to evergreen or live oaks, particularly the larger trees, such as coast live oak (*Quercus agrifolia*). Other local species that are evergreen include canyon live oak (*Q. chrysolepis*), scrub oak (*Q. berberidifolia, Q. dumosa*), and interior live oak (*Q. wislizenii*). "*Roble*" refers to deciduous oaks, such as valley oak (*Q. lobata*), blue oak (*Q. douglasii*), and black oak (*Q. kelloggii*). These two main categories may be further broken down into *encinitos* and *roblecitos*, diminutive terms which tend to be applied to smaller, shrublike species or individuals. Harrington's Chumash consultants did not specifically mention the evergreen tanbark oak (*Lithocarpus densiflorus*) but presumably thought of it as an *encino*.

Chumashan language terms distinguish at least four types of oaks, categories that seem to be somewhat more inclusive than botanical species. The various types of oaks could be used interchangeably for some purposes, but for certain uses one kind was preferred over the others. Insofar as these can be determined, the distinctive uses are listed separately under each species.

ACORNS

(Sp) Bellota

(B, I) 'ikhpanɨsh; shɨpɨtɨsh (acorn mush, soup)
(Cr) mɨsɨ; tluyash (acorn mush, soup)
(O) tpɨtɨ
(P) khakhpanɨsh
(V) 'ikhpanɨsh; shipitish (acorn mush, soup)

Acorns—an abundant, reliable, and storable resource—were the single most important plant food of the Chumash. Acorn use began among the ancestors of the Chumash at least five thousand years ago and continued until the early twentieth century. Although people often expressed preferences for some species of acorns over others, similar methods of harvesting, storage, and preparation were used for all.

Some acorns fell of their own accord, but one might also use a long pole to knock them from the trees. People picked the acorns up from the ground and placed them into net bags hung around the neck. They also used these gathering bags in collecting *islay (Prunus ilicifolia).* The bags were about the size of flour sacks, made in openwork mesh of Indian-hemp (*Apocynum*) fiber cordage or willow bark (*Salix*) with a wooden ring at the mouth.

After gathering the acorns, the Chumash spread them out on mats and dried them in the sun for fifteen or twenty days. It was also possible to dry acorns in temporary granaries outdoors. The granary was about three feet square and six feet high, constructed on a platform a foot or two above the ground. A plant called *'u'waw* (unidentified) was placed inside the granary to line the floor, walls, and top so that the acorns could dry without being attacked by pests. Large baskets might also serve as outdoor granaries. These they set on a platform, filled with acorns, and covered with inverted baskets, tule stems, and mats at night, being sure to remove the covering during the day to permit ventilation. The Chumash used these granaries only for drying, not for long-term storage, and they used them only in dry weather. Acorns do not keep well if left in the shell or outdoors for a long period of time.

The partially dried acorns were then shelled. The Chumash would place each one, tip down, in a small hole in a rock. Sometimes several holes had been bored in a single rock for more efficient work. They struck each acorn once or twice on the base with a hammerstone to crack it open, then set it aside. After having cracked a quantity of acorns, they rubbed handfuls at a time between their hands to remove the shells. The acorns were then further dried in the sun for another ten days to loosen the brown skin that adhered to the seed. The Chumash believed this skin

ACORNS FOR SUPPER

In a typical scene three hundred years ago, a Chumash mother rocks her baby's cradle with one foot while preparing her family's dinner. She pounds acorns into flour with a pestle in a bottomless basket fastened to a stone slab with natural tar. After washing the flour to remove the bitter tannin, she will cook it with heated stones in a basket. The family will eat the thick acorn soup with meat or fish.

served as protection from moths while the acorns were still on the tree. They tossed the acorns in a winnowing basket to remove the skin. By this time, the acorns were thoroughly dry and were placed into large baskets two and a half feet in diameter and two feet high for long-term storage inside people's houses. In mission times, the Indian adobe houses had lofts six or eight feet above the floor where people continued to keep these baskets of acorns and other wild seeds they had gathered.

A thriving trade in dry, shelled acorns between different parts of Chumash territory existed well into historic times. For this purpose, women's basketry hats, being fairly uniform in size, functioned as the standard unit for measuring volume. Acorns were less valuable than other seeds: one hatful of *islay* was worth two of acorns, and it took five hatfuls of acorns to buy one of chia. All these foods were traded between the mainland coast, interior, and offshore islands. It may be that the genetic diversity of oaks seen in the Chumash region today is at least partly a result of this aboriginal exchange network (Kevin Nixon, personal communication 1982; Timbrook 1993:57).

The Chumash ate acorns principally in the form of cooked mush or soup, called *shɨpɨtɨsh* in Chumashan languages, or *atole* in Spanish. To prepare the mush, they first pulverized dried acorns into flour by pounding them with a pestle in a stone mortar, using a vertical motion. Generally women did this, but sometimes men also did. Harrington's Chumash consultants did not mention sifting the flour and further pounding the coarser particles, but Kitanemuk people he interviewed described such a process at great length (Hudson and Blackburn 1983:134–137). Most California Indians did sift acorn meal.

Next, the acorn meal was leached in one of several ways to remove the bitter tannic acid. A simple method was to place the meal in a stone *olla* or wooden bowl and stir in fresh water, let it settle for several hours, and pour off the water. This process was repeated one to three times more until the bitter taste was gone. The Chumash generally used this method if they had only a little meal to prepare.

They also leached acorn meal in a basketry leaching basin, a twined openwork tray specially made for the purpose from whole stems of spiny rush (*Juncus acutus*) with a reinforced rim. They spread the meal over the basket in a layer about an inch and a half thick or, according to another version, around the edges in a spiral form. They placed the leaching basket on some rocks to hold it stable about six inches off the ground and slowly poured water through the fingers over the meal until the basket was full. Then they covered it with an inverted tray and allowed it to drain. They poured on more water a second and a third time if necessary, according to the amount of meal to be treated and the bitterness of the acorns. The Chumash said that too much leaching would make the mush insipid.

One could construct a leaching basin in clean sand if no leaching basket was available, but this was not traditional for the Chumash. In historic times, they would line a sand basin with a cloth before putting in the acorn meal. Otherwise the process was much like using basketry leaching trays. The Chumash did not leach whole acorns by burying them or by letting water run through them in a creek, methods practiced by some other peoples.

The most common way to cook acorn mush was in baskets using heated stones. While other peoples used rounded cobbles of certain igneous rocks for this purpose, Chumash boiling stones were slabs of soapstone in various shapes—Fernando Librado saw ones that were semicircular, square, or triangular—but they always had a hole near one end so they could be handled with a hooked stick. These stones were placed in the fire until they were quite hot. Meanwhile, the cook filled a large cooking basket with cold water and stirred in the leached acorn meal. When it was well mixed, she pulled several stones out of the fire and put them into the basket. The liquid would begin to boil right away. To ensure even cooking and to keep the hot stones from scorching the basket, it was important to keep stirring the mush with a wooden paddle. Eventually the stones would cool and the boiling would cease, and the stones would be fished out and placed on a mat. The boy who had gathered the wood for the fire got one of these stones to lick as his reward, but the laziest boy in the group was required to stir the mush while it cooked. An experienced mush maker touched her fingers to the mush on the stirring stick to tell when it was done.

They never added salt or other seasoning to the acorn mush. They scooped it from the bowl with their fingers or with a mussel-shell spoon and ate it with roasted meat, fish, shellfish, or seed flours. As María Solares said, "Whatever one has, one eats with acorn mush." If mush was left over, they allowed it to congeal in a wooden bowl and cut it into slices for eating. Usually, however, they only prepared the amount of acorn mush that could be consumed in two or three days, since it would go bad if allowed to sit longer than that. Very thick acorn mush was also prepared into lumps or balls called *kholowush* (I). According to Harrington's culture element checklist (1942:8), the Chumash supposedly made acorn bread in earth ovens as some other California peoples did, but no information about its preparation has turned up in his notes.

Individuals who were facing life crises or suffering ill health, as well as those who were about to take *Datura* or had recently done so, were required to follow a *sakhtin*, or special diet. Meat, grease, and salt were forbidden under this regimen, but acorn mush was nearly always permitted. Doctors prescribed a light broth of clams mixed into thin acorn gruel especially for sick people, who would often take this after purging with sea water. When a girl reached the age of first menstruation she drank only hot water, ate only acorn mush, and could scratch herself only with a piece of abalone shell.

Chumash consultants agreed that live-oak acorns (*Q. agrifolia*) were the best tasting and made the best mush. Black oak (*Q. kelloggii*), the species preferred by most California Indian peoples, is uncommon in the Chumash region and has not been identified from the information available. The Chumash considered acorns of valley oak (*Q. lobata*) and scrub oak (*Q. berberidifolia, Q. dumosa*) to be poor food, especially for mush, and generally avoided them. The acorns of a special type of live oak, as yet unidentified, were mild and sweet enough to be eaten raw; Harrington's consultants said these were nearly as small as pine nuts, with deep yellow shells and reddish caps. It is uncertain whether the Chumash utilized the acornlike seeds of tanbark oak (*Lithocarpus densiflorus*), which were prized by native peoples elsewhere in the state.

Nutritional analysis of dried, unleached acorns (Wolf 1945) suggests that preferences may have been based on fat content, those with more fat being regarded as better tasting. One early author suggested that the Chumash took laxative herbs to counteract the effects of an acorn diet (Bard 1894:5–6), but in fact the varieties usually consumed were sufficiently oil-rich to make such treatment undesirable (Wolf 1945:39).

The Chumash had a number of other, nonfood uses for acorns. They pierced whole acorns, still in the shell, near the tip and strung them into necklaces. They baited stone deadfall traps with acorns to catch ground squirrels. They rubbed raw acorn meal into their hair to make it grow well again after it had been trimmed by singeing. They chewed acorns and smeared this paste on the face to prevent sunburn. Various toys, including buzzers and a cup-and-ball game, were made from acorns (Harrington 1942:26, 27). To dye *Juncus* stalks black for basketry, the reeds were soaked in water with pounded acorns and iron (Blackburn 1963:144). The Chumash also made important religious offerings of acorns at the winter solstice and many other occasions throughout the year, and they made sure to store enough of the crop to supply future offerings in case the harvest was poor the next year (Blackburn 1963:145; Hudson et al. 1977:11, 45, 55).

Fagaceae — Oak Family

Quercus agrifolia NEE — **COAST LIVE OAK**

(B, I) ku'w — (Sp) Encino
(Cr) kuwu
(O) tsuwu'
(P) 'aku'w
(V) kuw

Coast live oak, which is an evergreen oak, or *encino*, is clearly identified with the Chumash category *kuw* or *ku'w*. Preparation of acorns for food is discussed in the previous section. Harrington's Chumash consultants expressed a definite preference for acorns of this species over others, despite the fact that they are quite bitter and require at least two or three leachings to make them edible. Dried but unleached acorns of this species contain 4.4 percent protein, 20.4 percent fat, and 52.7 percent carbohydrate (Wolf 1945:39). A single tree can produce several hundred pounds of acorns in a good year.

The Chumash observed that live-oak wood rotted quickly in the ground, so they did not use it for construction. They did find it valuable as fuelwood and for manufactured items. Stirring paddles made from *encino* wood were used in cooking mush. *Encino* was not the preferred wood for bow-making, as it was rather heavy, but very fine bows without a sinew backing could be made from the new shoots. These were bent to shape after being heated in a fire or with hot water.

The Chumash used this same method to shape oak shoots into the hoop for the hoop-and-pole game (Hudson et al. 1977:47). To play this game of skill, they rolled the hoop along the ground and contestants threw a lance, six or seven feet long and also made of oak, through it (Henshaw 1955:154). They also sometimes made live oak wood into game balls.

Chumash baby cradles were made on a forked framework with slats or crosspieces of *encino* twigs tied to it. One might instead heat the slats to bend them slightly and insert them through holes drilled in the frame. This made the cradle somewhat concave, but the mattress padding inside evened it out.

The Chumash made wooden bowls of various woods (see *Alnus*, *Platanus*, *Umbellularia*). Two of the five known extant Chumash wooden vessels—one a bowl, the other an *olla* or jar—are thought to be of oak (Hudson and Blackburn 1983:80, 249). Such containers were used for serving food and for storage. The Chumash may also have made mortars and boxes of oak. For a description of Chumash woodworking technology, see Hudson 1977.

Several Chumash consultants said that oak bark was the favored kind of firewood. One could knock it off the tree with any kind of instrument, and at the right time of year it would fall off in large chunks at a single blow. *Encino* bark, when burned, produced long-lasting coals that could be kept buried overnight and stay hot enough to kindle a fire the next morning. This method was easier and was used more often than making a new fire with firesticks. The Chumash also toasted seeds by tossing them with hot coals of *encino* bark on a basketry tray.

The Santa Cruz Island Chumash wore their hair in bangs, which they kept trimmed by singeing with a burning oak coal. When mourning the death of a relative, Chumash people would have their hair cut short with a knife and then singed to an even length, holding the hair between the fingers and applying the coal to the projecting ends. To remove the burned smell, they rubbed the singed ends with ground charcoal or with ground raw acorn meal, which they believed would make the hair grow well again.

Some consultants said that the red-colored fresh inner bark of *encino* was used for dyeing hides, particularly cowhide in historic times, but they did not describe the process. They may have been referring to the common European method of tanning hides by soaking in an oak-bark solution, which colors the leather as well as softening it.

Both oak bark and oak galls had medicinal uses. The Chumash burned the fresh green bark down to charcoal, mixed it with water, allowed it to sit overnight, and then drank the liquid for indigestion and bowel trouble. They said this "charcoal soup" was a purgative that was effective in curing even when other methods failed. In Hispanic times, a slightly different recipe for the same treatment was to pound oak bark with water, add some salt, and let it stand overnight out in the open before drinking it. The Chumash also mentioned soaking live-oak bark in water and drinking it to treat pustules and boils.

The juice from fresh oak galls was applied to any pustule or wound, and charred oak galls were considered a good remedy for hemorrhoids. Missionaries remarked that the native people used oak bark to treat every sort of wound (Geiger and Meighan 1976:72). Oak juice, flowing most heavily at the time of the new moon, was also thought to be of medicinal value. The high tannin content of the oak bark and galls might make them an effective remedy where astringent action was needed. Nowadays oak bark is chewed in order to tighten and strengthen the teeth (Weyrauch 1982:13).

Fagaceae — Oak Family

Quercus berberidifolia LIEBM. — **SCRUB OAK**

Quercus dumosa NUTT. var. *dumosa* — **NUTTALL'S SCRUB OAK**

(B, I, V) mɨs — (Sp) Encino Chino, Encinito

Encino chino, or "curly live oak," was a small, gnarled live oak that the Chumash considered a separate species. Lucrecia García collected a specimen of this type of oak and labeled it with Chumash and Spanish names. Botanists identified this specimen as *Quercus dumosa* var. *dumosa*. One oak specialist commented that the specimen probably came from María Ygnacio Creek, without knowing that Lucrecia García had a ranch there (Kevin Nixon, personal communication

1982). Some plants formerly classed as *Q. dumosa* have now been assigned to *Q. berberidifolia* and other closely related species that are difficult to distinguish visually (Smith 1998:255–256), although their distributions differ in relative elevation and distance from the coast. It is doubtful that the Chumash made any distinction between these very similar scrub oaks.

Lucrecia García said that people did not make soup from *mɨs* acorns. Her mother, Luisa Ygnacio, told Harrington that people did eat them but considered it poor food. This kind of oak was used to make bows, arrow foreshafts, and thatching needles. It is very flexible and strong, Simplicio Pico said, and a stick of it could be bent double without breaking.

Fagaceae — Oak Family

Quercus douglasii HOOK. & ARN. — **BLUE OAK**

(I) tushqun — (Sp) Roble Chiquito, Roblecito

(P) mish'kata

This type of oak was described as smaller than valley oak, growing more short and squatty, and with very hard, dense wood. No specimens of it are extant in the Harrington collection. María Solares had seen these growing in the mountains southwest of Tejon. She said that some fine wood bowls were made of the burls of *tushqun*. Two of the extant Chumash wooden bowls are made of an unidentified species of oak (Hudson 1977:9).

Fagaceae — Oak Family

Quercus lobata NEE — **VALLEY OAK**

(B, Cr, I, O) ta' — (Sp) Roble

(V) ta

Oaks that lose all their leaves seasonally are known by the Spanish term "*roble.*" Specimens collected by Lucrecia García indicate that the Chumash taxon *ta'* includes both *Quercus lobata* and the natural hybrid complex some have recognized as *Q. dumosa* var. *kinselae*, which is a small deciduous oak found in the foothills near Santa Barbara (Smith 1998:256). Both of these have distinctively shaped, lobed leaves. Valley oak and hybrid forms also occur on Santa Cruz Island (Junak et al. 1995).

Although some Chumash descendants have said that valley oak acorns are desirable because they are large and not very bitter (Weyrauch

1982:13), Harrington's consultants considered them inferior. The flavor was described as insipid, perhaps due to the low fat content of this species: about 5.5 percent in whole, unleached acorns (Wolf 1945:30). *Roble* acorns were called *'asaqa* in Ineseño Chumash, the same as the name of Zaca Ranch. Another type of *roble* with small acorns that were sweet-tasting was called *'aqwachmu'* (I).

The Chumash said that *ta'* was not good for firewood, since it just made ashes rather than coals. In historic times, wagon axles were made from *roble* wood.

Rhamnaceae — Buckthorn Family

Rhamnus californica ESCHSCH. — **COFFEEBERRY**

(B, I) puq' — (Sp) Yerba del Oso

(V) chatɨshwɨ 'ikhus

The Ventureño Chumash and Spanish names for this plant have the same meaning, "bear's medicine," possibly in the sense of a talisman (*'atishwɨn*) conferred by a personal spirit guide or "dream helper." Several Chumash consultants commented that *yerba del oso* berries were food of the bear but they were poisonous to humans, and eating them would make a person crazy. Rosario Cooper believed it would paralyze the jaws.

The Chumash did use *yerba del oso* for medicinal purposes, however. They rubbed the leaves on the skin as a remedy for rheumatism. To treat poison-oak rash, people would bathe in water in which the leaves of this plant had been boiled (Birabent n.d.; Gardner 1965:298; Brand and Townsend 1978:43). Most evidence indicates that Chumash people have not always been troubled by poison-oak (see *Toxicodendron*).

The Chumash also boiled coffeeberry bark to make a laxative tea, according to Fernando Librado, and in the 1960s they were drinking this decoction for stomach gas (Gardner 1965:298). Although *Rhamnus* bark tea was held by some authors to be beneficial for "the constipating effects of an acorn diet" (Bard 1894:5–6), others suggest that the preferred acorn species, which have relatively high oil content, would themselves act as laxatives (Wolf 1945:39). *Cáscara sagrada (Rhamnus purshiana)*, a similar species found in northern California, was at one time harvested commercially and made into an over-the-counter laxative preparation called Cascara. In some areas, *R. californica* is also locally known as *cáscara sagrada* (Chesnut 1902:368–369; Saunders 1920:195–198).

Other California peoples made similar use of *yerba del oso*, and its medicinal value was widely recognized (Strike 1994:131–132; Mead 2003:347–349). The Ohlone decocted the bark for use as a laxative and purgative, made a poison-oak remedy from its leaves, and reportedly also ate the berries raw (Bocek 1984:250).

Rhus laurina: See *Malosma laurina;* see also *Rhus integrifolia, Rhus ovata*

Anacardiaceae — Sumac Family

Rhus integrifolia (NUTT.) BREWER & S. WATSON — **LEMONADEBERRY**

Rhus ovata S. WATSON — **SUGAR BUSH**

(B, I) walqaqsh — (Sp) Mangle Menor

(V) shtoyho'os

The Barbareño and Ineseño Chumash lumped these two species of *Rhus* together into the same folk taxonomic category as laurel sumac

LEMONADEBERRY
Rhus integrifolia

(*Malosma*, formerly *Rhus laurina*). The Ventureño separated them into two kinds, a larger and a smaller, with the smaller, or "*mangle menor*," category consisting of *Rhus integrifolia* and *R. ovata*. The Spanish common name "*mangle*" was bestowed because of a fancied resemblance between these shrubs and mangroves found in tropical wetlands.

Fruit of sugar bush and lemonadeberry may have been eaten in the same way as berries of laurel sumac: pounded, dried in the sun, and eaten without cooking. Harrington's Chumash consultants did not mention soaking any of these fruits in water to make a beverage. Several other California people did so, including the Cahuilla, who also ate the berries of *Rhus* spp. fresh or dried, or made into mush (Bean and Saubel 1972:131–132).

Anacardiaceae — Sumac Family

Rhus trilobata TORREY & A. GRAY — **THREE-LEAVED SUMAC**

(B, I) shu'nay — (Sp) Chiquihuite

(V) shuna'y

Most Indian people today prefer the English names sumac, sourberry, or basket bush instead of the once-popular term "squawbush." Harrington's Chumash consultants often called this plant by the Spanish term "*chiquihuite*," which actually refers to a twined openwork utility basket made from shoots of sumac or willow. All these individuals clearly distinguished *shuna'y* and willow as two different plants, and willow was given its own Chumash names. A previous writer's identification of *shuna'y* as willow was mistaken (Craig 1967:90).

The Chumash used the straight, woody stems of *Rhus trilobata* in making several kinds of baskets. One of the most important was the seedbeater, a type of basket much like a tennis racket in size and shape. The Chumash version had a fan-like blade about a foot in diameter and a handle six to twelve inches long. This implement was used to collect small seeds, particularly chia (*Salvia columbariae*), by knocking them loose with a sweeping motion. Fernando Librado, who had last seen seedbeaters in use in Ventura about 1849, described one method of making them. Bend a long, unpeeled stem of sumac into a curve and tie the two ends together to form a handle. Place a second piece lengthwise onto this, so that one end forms part of the handle and the other extends to the curve in the middle of the top, and tie that in place. Lash three crosspieces onto this framework at the center and sides. Then place small *shuna'y* stems close together side by side on this frame, making a notch at the right spot, and tie them in place. From Fernando Librado's description, this type of seedbeater seems to have been merely tied together, not woven or twined.

Also made of sumac shoots were the *chiquihuite* baskets. Though they may have had antecedents in open-twined utility baskets, this form was possibly a historic phenomenon rather than indigenous with the Chumash. María Solares said there were no *chiquihuites* in the old days before the missions. Consultants described loads of these baskets being made and taken for sale as far away as Los Angeles. The *chiquihuite* was done in wicker weave, not twining, of whole or split sumac stems. The maker started by laying out six splints side by side and weaving six more across them at right angles. Weaving the sides proceeded with another strip. They usually peeled the sumac stems, but they left the bark on some of them for making brown designs. They wrapped the rims with walnut bark (Craig 1967:112). These baskets could also be made of willow shoots or of *carrizo (Arundo*, *Leymus*, or *Phragmites)* stems. An example of a wicker-woven basket that belonged to Candelaria Valenzuela is in the Southwest Museum collection.

María Solares said that water bottle baskets, tarred inside, could be made of *Juncus* or of sumac. There is one known surviving example of a sumac water bottle from Chumash territory, diagonally twined with split sumac on a warp of peeled shoots, but it resembles one found in the Gabrielino area and may be a trade piece (Dawson and Deetz 1965:201–202). Native peoples in the Great Basin and Southwest made that type of water bottle.

Chumash coiled basketry sometimes incorporated split, peeled sumac stems as white sewing material. Straight shoots up to five feet long with no knots were most desirable for this purpose. It was customary to split each stem lengthwise into four quarters, peel the bark off and clean the pith from each quarter. Each of the four strips was then trimmed and flattened by scraping it with a sharp piece of clamshell. This process also made the strips flexible enough to sew with, using a bone awl in the same manner as one would sew with *Juncus*. The Chumash stored the split and trimmed strips in bundles an inch thick, tied with three ties. It took several such bundles to make a large basket.

SUMAC
Rhus trilobata

Harrington's consultants described several types of baskets that were sewn with *shuna'y*. These included the bucket-shaped burden basket (*wo'ni*), the trinket basket (*kh'omho*), the boiling basket (*watik*), the shallow basin (*'ayu'hat*), and the *'epsu*, or basket hat (Craig 1967:90, 95). The toasting tray (*yep*) was made entirely of sumac and coated with tar inside. Seeds were tossed with hot coals in this tray, but somehow the tar did not melt and stick to them.

Fernando Librado and María Solares said the Chumash made the core of certain coiled baskets from this plant, but sumac foundations have not been observed in extant specimens. Many baskets with white backgrounds or design elements of sumac do exist. The Chumash also patched old, worn baskets by gluing on fragments of other baskets and sewing them with strands of sumac (Dawson and Deetz 1965:203).

Any item of dance regalia that required a rigid framework was also made with sumac shoots. The Fox Dance headdress had a dome-shaped frame with three horizontal rings tied onto two vertical, arched members. Over this armature they wove tule and covered the whole thing with fox skin, sometimes adding a real fox head, and always a

SUMMER'S HARVEST

A Chumash mother and daughter gather wild grass seeds along the coastal plain, on a midsummer day five hundred years ago. Dried seeds can be stored for months, then toasted and ground for eating. In the distance, old women have finished harvesting and are burning the dry grass to ensure a more plentiful crop for the next year. As the Chumash and their ancestors have known for centuries, edible wildflowers, grasses, and bulbs can grow much better after a fire.

long, weighted tail (Hudson and Blackburn 1985:195). The *tsukh*, or topknot headdress, had a ring-shaped frame of sumac which served as the base for the crown of split feathers.

On the gathering net for collecting *islay* fruit and acorns, the Chumash added a wooden ring of either sumac or willow to hold the net mouth open. They also sometimes bundled sumac stems for use as brooms.

Although the sour-sweet fruit of *Rhus trilobata* is edible and was utilized by some other California peoples (Mead 2003:153), the Chumash apparently did not make use of it.

Sumac was particularly important as a sewing material in coiled basketry among peoples who lived south of the Chumash. In central California, where the whole and split shoots were made into twined, scoop-shaped gathering baskets, this species was generally known as sourberry. It is one of the plants widely managed by burning to produce long, unbranched shoots for basketmaking (Anderson 2005).

Grossulariaceae — Gooseberry Family

Ribes spp. — **GOOSEBERRY**

(B, I) stɨmɨy — (Sp) Barburi, Baburi

(V) chtɨmɨy

Ribes speciosum PURSH — **FUCHSIA-FLOWERED GOOSEBERRY**

(B) stɨmɨy 'iwɨ — (Sp) Barburi, Baburi

(I) tsiqun

Ribes spp. — **CURRANT**

(B) sqa'yi'nu — (Sp) Barburi, Baburi

(I) sqayi'nu

The Chumash ate the fruits of several species of gooseberries and at least one of currants. Most people, the Chumash included, distinguish these two types of berry bushes from one another according to whether or not they have thorns: gooseberry plants have spiny branches and the fruit is often bristly, while currants lack spines. The following attributions are based on descriptions that Harrington's consultants gave of the plants, as well as on the specimens they collected.

Spiny *barburis* were called *stɨmɨy* (B, I) or *chtɨmɨy* (V). María Solares added that this was the big, fat kind; she said there was also a "little *barburi*" called *tsiqun* (I), apparently a smaller-fruited spiny kind. The "fat" kind, identified from specimens labeled *stimiy*, was bitter gooseberry (*Ribes amarum*). The "little" spiny kind may have included canyon gooseberry (*R. californicum*) and fuchsia-flowered gooseberry (*R. speciosum*).

Specimens of fuchsia-flowered gooseberry were labeled "*barburi de venado, stɨmɨy 'iwɨ*" (B); both the Spanish and Barbareño names mean "deer gooseberry." Lucrecia García said that people did not eat berries of this kind.

The not-thorny category was *sqayinu* (B, I), "smooth little *barburis*," which probably included chaparral currant (*R. malvaceum*). The Ventureño Chumash acknowledged the non-spiny variety, but no separate name was recorded for it in that language.

Many kinds of gooseberries and currants were eaten by native peoples throughout California. The leaves, fruits, and roots also had various medicinal uses (Strike 1994:133–134; Mead 2003:354–358).

FUCHSIA-FLOWERED GOOSEBERRY
Ribes speciosum

Brassicaceae | Mustard Family

Rorippa nasturtium-aquaticum (L.) HAYEK

WATERCRESS

(B) welu

(V) spe'ey he'so'o

(Sp) Berro

Botanists formerly believed that watercress was introduced from Europe (see Smith 1976:149), but most now consider it a California native. The Spanish explorer and missionary Juan Crespí found it growing in Price Canyon, San Luis Obispo County, as early as 1769 (Crespí 1927:183). The 1813 location of the second La Purísima Mission was in the Canyon of the Watercress—Cañada de los Berros. The Barbareño name *welu* is a loan word derived from the Spanish "*berro.*" The Ventureño name *spe'ei he'so'o* is actually a description, "flower of the water."

Chumash people ate watercress raw and sometimes boiled it to eat as cooked greens, and it is still eaten raw, plain, or in salads (Weyrauch 1982:5). Indians and settlers ate the plant raw for liver ailments (Birabent n.d.). Santa Ynez Reservation residents said that illness caused by drinking too much alcohol was treated by eating the plant raw or by drinking a tea made from it (Gardner 1965:299).

Other California peoples also ate the leaves raw or as cooked greens (Mead 2003:359). The Cahuilla ate the leaves and reportedly used the plant for liver ailments and low blood pressure (Bean and Saubel 1972:90). The medicinal use of watercress is widespread throughout the Southwest, where it is employed to treat heart and kidney ailments, tuberculosis, and influenza (Ford 1975:142–143).

Rosaceae | Rose Family

Rosa californica CHAM. & SCHLDL.

CALIFORNIA WILD ROSE

(B, I) washtiq'oliq'ol

(V) watiq'oniq'on

(Sp) Rosa de Castilla

Although Crespí called this plant "Castilian rose" in his 1769 journals and the Chumash also used that Spanish name for it, it is a native California species. The Chumash ate the fruit of the wild rose raw. They also pierced the fruits, or rose hips, and strung them for wearing as necklaces and earrings, especially by children.

This plant was regarded as a great remedy for ailments suffered by children. The Chumash used dried, powdered rose petals like talcum to

CALIFORNIA WILD ROSE
Rosa californica

relieve chafing and skin rash in babies. They made the petals into a tea which was administered for stomach pain or colic (Birabent n.d.). In recent decades, Chumash people have used these flowers in a medicinal wash for the eyes and for teething babies (Gardner 1965: 298).

Indian and Hispanic people in Santa Barbara continue to gather wild rose petals in late spring and dry them for later use. Rose petal tea is still used as a wash for sore eyes and is given to babies for colic, constipation, and difficulty in teething. Adults also use this tea, often mixed with other herbs, for *empacho* (blockage in the stomach); the petals may be mixed with egg white, put in a flannel cloth, and applied to the stomach to draw out fever and sickness (Weyrauch 1982:17). Similar uses are reported elsewhere in California (Strike 1994:135) and the Southwest (Ford 1975:290–291).

Rosaceae — Rose Family

Rubus ursinus CHAM. & SCHLDL. — **CALIFORNIA BLACKBERRY**

(B) tɨq'ɨtɨq' — (Sp) Mora (fruit), Salsamora (plant)

(I) tɨqɨtɨq

(V) tɨhɨ

"*Salsamora*" was probably the Chumash pronunciation of the Spanish common name "*zarzamora*," meaning the wild blackberry plant or vine. "*Mora*" refers to the fruit. The Chumash ate blackberries and generally considered them a delicacy, but they believed that those growing in a damp location could be unwholesome and potentially injurious. Wild raspberries, *R. leucodermis*, may have been eaten where available.

The northern Chumash and the Salinan boiled blackberry roots into a tea taken for diarrhea (Garriga 1978:25). The Ohlone also considered this the most effective remedy for dysentery (Bocek 1984:250). Many groups also utilized the berries for food, and the Luiseño stained wooden articles with blackberry juice (Sparkman 1908:232).

Polygonaceae — Buckwheat Family

Rumex crispus L. (Introduced) — **CURLY DOCK**

(B) tsukhat' — (Sp) Lengua de Buey

(I) ts'ukhat'

(V) 'alakhnipk'ɨsh

The Chumash used all parts of this introduced plant. They cooked the leaves as greens and ate the raw, peeled stem like celery. It had a sour taste, hence the plant's descriptive name in Ventureño, meaning "the flavor is sour." María Solares of Santa Ynez said the seeds were good for making into *qolowush*, a molded thick mush. According to Luisa Ygnacio, this food had gone out of use in Santa Barbara by 1914. Ash bread, *kapit'*, was also made of *lengua de buey* seeds, pounded up raw into a dough, molded into balls, wrapped in tule, and baked in hot ashes. The Chumash boiled the root of this plant to make a medicinal tea for stomach trouble.

The Spanish common name, "ox tongue," describes the shape of the leaf. Curly dock was introduced from Eurasia, but it is possible the native willow dock (*R. salicifolius*) was used in the same way.

Other California peoples ate both leaves and seeds of *Rumex crispus* and used the plant medicinally (Chesnut 1902:345–346; Bocek 1984:249). In mission times the root was used to treat diphtheria and syphilis (Garriga 1978:25, 40). It is recognized as having astringent and tonic properties (Jepson 1914:385).

Polygonaceae | Buckwheat Family

Rumex hymenosepalus TORREY

WILD RHUBARB, CANAIGRE

(Sp) Cañagria

(B) shaw

(I) sha'w

(V) 'alakhpɨy

Chumash people collected the tender shoots of *cañagria* in March and ate them boiled or roasted. The taste was sour but agreeable, and the Ventureño sometimes referred to the plant as *'alaqpɨi*, "sour." The Ineseño made a molded thick mush called *qolowush* from the seeds. A tea made by boiling the root had medicinal uses, particularly for liver trouble, sore throat, and dizziness (Garriga 1978:30, 33, 41).

Fernando Librado said that the large root of the *'alaqpɨi* plant was used to make a yellow dye for basketry materials. He saw people put mashed-up roots in water and soak split *Juncus* in this solution for several days.

Many native peoples in California and the Southwest made use of canaigre. The Cahuilla ate the stalks and used canaigre roots in tanning hides (Bean and Saubel 1972:134–135). This root has had a wide variety of medicinal uses, including treatment of loose teeth, sore throat, stomach trouble, and skin inflammations (Ford 1975:153–154). In mission times, canaigre root was reportedly made into a mustard-colored dye called "padre's gold" (Grae 1974:83).

Salicaceae — Willow Family

Salix spp. — **WILLOW** (general)

(no generic Chumash name known) — (Sp) Saus, Sauz

Specimens collected by Harrington's consultants indicate the Chumash recognized two basic kinds of willow; Chumash taxonomic categories may well be more inclusive than today's botanical species. Consultants often specified *wak* or *saus chino* (identified as red willow, *Salix laevigata*) for certain purposes, and *shtayɨt* or *khaw* (arroyo willow, *S. lasiolepis*) for others. In many other cases, however, they just referred to *saus* or willow without saying which kind was used.

This section covers those uses that, at least in the information recovered from Harrington's notes, were not specifically attributed to any particular type of willow. Uses for which a named category of willow was specified are listed separately (see *Salix laevigata* and *S. lasiolepis*). It is unclear where other local species, such as yellow willow (*S. lucida* subsp. *lasiandra*), may have fit in the Chumash classification scheme.

Willow was an important construction material, particularly for building houses. Spanish explorers in the sixteenth, seventeenth, and eighteenth centuries described Chumash houses as hemispherical, round, and dome-shaped like half an orange, and capable of housing forty to sixty people (Bolton 1925:30, 34, 90; Crespí 1927:24, 38, 158; Brown 1967:4; Fages 1937:24–25). The full text of these descriptions has been compiled by Hudson and Blackburn (1983:323–325). Photos of a reconstructed Chumash house built by Harrington's consultants for the 1924 Ventura County Fair (Hudson and Blackburn 1983: 334–337) show a more conical structure, with the vertical poles coming together at the top but not bent over far enough to make a dome. Fernando Librado said the Indian houses of the Lulapin, or people in the Point Mugu area, were conical and high enough to stand up in.

Librado said the house framework was made of about fourteen willow poles, each fourteen to eighteen feet long. The Chumash sometimes used sycamore (*Platanus*) or cottonwood (*Populus*) for this purpose, but they preferred willow. They did not burn the poles but always cut them down. It was best to get the poles in March when the bark was fresh, for the bark would be used in making the ties for the house thatching. They removed the leaves, twigs, and bark from the poles and charred the cut ends to retard decay. Post holes were dug in a circle, about the width of a person apart, leaving a wide place where the door was to be. They planted the bases of the poles in these holes, bent the tops over, and lashed them to the tops of opposite poles to form a dome.

Once the vertical poles were in place, the builders added inch-thick willow sticks as horizontal crosspieces on the outside of the frame, tying them in place with strips of willow bark wherever they crossed one of the uprights. This lattice-like framework was then covered with thatching of tule (*Scirpus*), cattail (*Typha*), bracken fern (*Pteridium*) or *carrizo* (*Leymus*). They put the thatch on in courses, starting at the bottom and working toward the top, tying each course in place with willow bark or with cordage. Willow bark was a very important lashing material, as well as being a significant source of fiber for making a variety of objects (see *S. laevigata*).

After a long while, the base of one or more of the poles might decay in the ground. If the upper structure was still sound, it was possible to reinforce the base of the decayed pole. They inserted a new willow post into a hole dug in the ground next to the decayed base and lashed it to the above-ground part of the original pole.

Inside the house there were sleeping platforms made by setting four forked willow poles in the ground. Two long side-pieces and several shorter crosspieces were tied on with cordage of Indian-hemp (*Apocynum*), making a platform about three feet high which was then covered with layers of tule mats. The Chumash also built raised platforms inside the house to keep large storage baskets of seeds dry and safe from rodents. These storage platforms were about three feet high, three feet wide, and five feet long, with willow rods placed close together and tied in place with willow bark. Willow storage platforms were made outdoors as well.

Chumash people employed willow poles in building other kinds of structures, too. Each village had a ceremonial enclosure, or *siliyɨk*, a semicircular area bounded by willow poles sunk into the ground at six-foot intervals and interwoven with willow twigs to form a wall about five or six feet high. At the last festival held in Saticoy in 1869, the *siliyɨk* was twenty-five feet long, and straight rather than semicircular (Hudson et al. 1977:91).

A *ramada*—a roofed structure open on all sides or partially enclosed—served as a shady work area in warm weather. At the four corners were forked willow poles, and the roof and walls were also of willow poles. Fernando Librado said that the *ramada* was not an indigenous Chumash structure, but it did become common in Indian communities in historic times (Hudson and Blackburn 1983:350).

The ladder for the sweathouse was a notched pole tied to one of the roof beams. Willow was the preferred wood for this purpose. One climbed down this ladder from the roof opening into the sweathouse.

Dugout canoes were hewn from a single large tree trunk, preferably willow but cottonwood might also be used. The local willow with the thickest trunk, yellow willow (*Salix lucida* subsp. *lasiandra*), would probably be most suitable for dugouts. This type of watercraft was uncommon among the Chumash in historic times, and it appears to have been restricted to use in estuaries. Few details are available about its manufacture. For example, Fernando Librado, who had seen dugouts made and used, was uncertain whether or not fire was used to hollow out the interior of the dugout, or whether the interior was shaped before or after the exterior (see Hudson, Timbrook, and Rempe 1978:31–36). During construction of the more famous plank canoe, or *tomol*, willow forks supported each end of the vessel.

From a study of extant baskets, Dawson and Deetz (1965:203) found that the use of willow as a basketry material was rare among the Chumash, being confined to peeled shoots as twining warp. Harrington's consultants mention a number of kinds of baskets—primarily simple, utilitarian items—that were made with willow (see also *Salix laevigata* for Chumash basket types made from *wak*, or *saus chino*). Native weavers throughout California generally regarded narrow-leaved willow (*S. exigua*, including the sandbar willow, which was formerly considered a different species known as *S. hindsiana*) as the best source of shoots for basketry (Tulloch 1990). Vouchers of that species were not found in Harrington's Chumash plant collections, but it is possible that *S. exigua* might be the "*saucillo*" that Fernando Librado said could be used for making *chiquihuite* baskets.

Seedbeaters were usually made from three-leaved sumac (*Rhus trilobata*), but willow might also be used. One type of seedbeater had a frame of willow with five or six twigs of the same material inside the frame and woven across with string. The Chumash would gather chia (*Salvia columbariae*) by knocking the heads of the plant with this seedbeater, so that the seeds fell into a large basket suspended from the waist.

One type of Chumash baby cradle was Y-shaped, with a forked willow frame and short willow crosspieces which fit through holes in the sides of the frame. Candelaria Valenzuela said that fresh, green willow was preferred for making the cradle since it was easier to work before it dried. A mattress and sunshade of tule were tied to this frame. (See Craig 1967 and Hudson and Blackburn 1982:316–323 for more information on Chumash cradles.)

Certain types of gathering and carrying baskets and nets, which received heavy use, had their rims reinforced with rods of willow lashed on with strips of willow bark. Harrington also described a leaching tray for acorn meal that was quickly made by twisting and weaving willow twigs together (letter to J. W. Fewkes, Nov. 9, 1925, sent from Santa Barbara), but this may not have been a Chumash technique.

In Hispanic times, the Ventureño made a sort of storage bin of willow twigs. Fernando Librado described these as three feet long and two feet wide, with the twigs coming together at dovetailed corners. He said these bins—intended for storage of raw Mexican sugar cane or *panocha*, lumps of brown sugar—were about eighteen inches high, but there was also a shorter version that was more portable.

The Chumash made a variety of tools from willow wood (see also *S. laevigata* and *S. lasiolepis*). One of these was the thatching needle of willow or other wood, about five feet long and an inch thick, with one end worked to a point and perforated. In building a house, they attached cordage to this thatching needle, then passed it through the layers of tule or other thatching materials and around the horizontal members of the framework. Firesticks might be made of willow, but *guatamote (Baccharis salicifolia)* was more commonly used. Historically, forked willow poles were used as pitchforks in loading grain into wagons.

Musical instruments and dance regalia were also made from willow. The whirring sound of the bullroarer announced ceremonial activities and accompanied certain dances. This instrument was a thin slat of willow or elderberry (*Sambucus*) wood, two feet long and painted red. The player created its distinctive sound by whirling the instrument overhead by means of a string attached to one end. Dance regalia made with willow included the feathered wands dancers held during the Seaweed, Barracuda, and Beaver Dances. Willow was not considered suitable for making the feathered poles erected at shrines.

Willow, prized as firewood, was the only fuel used in sweathouses. The Chumash usually took a sweatbath around five o'clock in the evening. Participation was voluntary for both men and women as well as for young people, but children were not permitted. The fire was made with thin, dry willow twigs—only dead wood, never cut from a live tree—to make a big blaze. Four bundles of wood, each a foot and a half in diameter, were added to the fire one at a time as the previous one burned down. After the fourth bundle had been burned, everyone left the sweathouse and jumped into cold water to bathe. They then returned to the sweathouse, without the fire, to temper the blood.

Medicinal treatments employed willow bark, branches, leaves, and roots. See *S. lasiolepis* for specific examples. Willow bark was said to be used in tanning or dyeing deer skins.

A willow stick that had been cut by a beaver was thought to have the power to bring water. The Chumash would treat the stick with *'ayip* (a ritually powerful substance made from alum) and then plant it in the ground to create a permanent spring of water. María Solares said she had encountered this practice among the Indians near Tehachapi (possibly the Kawaiisu), but the Chumash apparently shared the same belief.

Salicaceae — Willow Family

Salix laevigata BEBB — **RED WILLOW**

(B, Cr, I, V) wak — (Sp) Saus Chino

Harrington's consultants specified this type of willow for several purposes. It was botanically identified from specimens collected by Lucrecia García. The Spanish common name means "curly willow."

Some of the willow baskets previously described, particularly the twined openwork baskets called *chiquihuites*, may have been made from this species. Fernando Librado recalled a time when the Ventureño Chumash took many *chiquihuites*—baskets made of *saus chino* or *saucillo* willow shoots—to Los Angeles and sold them for a dollar apiece. María Solares said that the seedbeater was made of very fine sticks of *saus chino*. In central California, many weavers considered *S. laevigata* too large-pithed, branching, and brittle to be ideal for basketmaking (Tulloch 1990).

The Chumash processed the bark of *saus chino* into a soft fiber from which they made a variety of articles. All uses that Harrington's consultants described for willow-bark fibers are listed in this section.

In March, when the sap was flowing, it was easy to pull the bark from the trees in long strips, starting from a horizontal cut made at the lower end. Within a day after collecting willow bark, people pounded or mashed it (on one occasion Fernando Librado said they put it into hot water, but on another he denied they ever soaked it) and allowed it to dry. Then they removed the outer bark by rubbing the material in the hands, holding a bunch in one hand and revolving the other clenched hand on top. This process softened the bast into usable fiber.

They twisted the shredded *saus chino* fiber into two-ply cordage of various thicknesses, or used it without twisting, to make several kinds of garments. Three strands of *saus chino* bark were woven or braided to make at least two kinds of belts. Hunters wore a game-carrying belt that was about three-quarters of an inch wide and long enough to go around the waist two or three times. Men also wore a three-inch-wide belt with their loincloths. To make sandals, some people took strips of *saus chino* bark, doubled them repeatedly, laid them side by side, and sewed them together. They sewed a grass covering to the top and attached strings as straps (Hudson and Blackburn 1985:97). The Fox Dance headdress had a long tail made from braided willow bark or tule and weighted with a stone on the end.

Chumash women of the poorer classes wore skirts of tule (*Scirpus*) or the shredded inner bark of willow or cottonwood (*Populus*). The fiber skirt consisted of a front apron and an optional back apron (Harrington 1942:19), or sometimes it was large enough to go around the wearer's body and tie at one side. It could be made either of lengths of loosely twisted two-ply *saus chino* cordage about three-sixteenths of an inch thick, or of softened strips of *saus chino* bark left untwisted. The strands hung from a wide woven belt of *saus chino* bark and were twined or woven at the top with cordage of the same material. They hung loose like fringe nearly to the knees.

Willow bark was also woven into bags and nets, particularly those used for gathering *islay* fruit (*Prunus ilicifolia*) or acorns. One kind of fishnet, six or eight feet long and about eighteen inches in diameter at the mouth, was made of two-ply willow cordage one-eighth of an inch thick. Hunting slings, tumplines, carrying rings, tarring brushes, arrow quivers, and saddle pads were also made from willow bark fibers.

Various consultants also reported the use of *saus chino* wood to make mush-stirring sticks, spoons, digging sticks, and possibly wooden bowls. The Chumash often weighted their digging sticks with a perforated stone fixed about a foot above the pointed end. A game ball used in one version of the hoop-and-pole game was made from wood of *saus chino*, live oak (*Quercus agrifolia*), or stone; the ball was thrown or rolled through a hoop of straw.

Salicaceae — Willow Family

Salix lasiolepis BENTH. — **ARROYO WILLOW**

(B, I, P) shtayɨt — (Sp) Saus, Sauz

(Cr, V) khaw

(O) tsa'

The kind of willow that Harrington's Chumash consultants called *shtayɨt* or *khaw* was identified from specimens of *S. lasiolepis*, but there was scant information in his notes about use of this type. The paucity of data may indicate that this species was not much used, so only those few uses securely attributed to this type of willow are listed here. But it is also possible that many of the uses ascribed to an unspecified type of willow (given under *Salix* spp. above) actually employed this species. Arroyo willow was favored by some California basketmakers for its large, long shoots (Tulloch 1990).

Several Chumashan place names incorporate the name of this species. One of these is on Santa Cruz Island at the location of the present Willows Canyon. Others include *kakhaw*, near Ventura, *kashtayɨt*, west of Gaviota, and *mishtayɨt*, near Arroyo Burro in Santa Barbara (Applegate 1975b:31, 32, 36, 38).

Mary Yee had heard that they cut leafy branches of *shtayɨt* and spread them out as a bed for a feverish person to lie on in order to cool the body. She added that the early people already knew that it could harm you to drink cold water when you are very hot.

Medicinal uses of willow in the treatment of fever and inflammations have been reported by other writers and may also refer to this species. Small willow roots were used to treat fever in the Mexican colonial era (Bard 1894:16), and it is possible the colonists may have adopted this remedy from the Chumash. Indians and settlers prepared a strong decoction of willow bark and leaves to bathe hemorrhoids (Birabent n.d.). Early families made willow into a gargle for sore throat (Benefield 1951:24). Willow-bark tea has continued to be used for fever and malaria, and the bark is chewed to strengthen the teeth (Weyrauch 1982:20).

In 1978, an elderly Chumash consultant commented that a tea good for the blood could be made from the bark of a willow tree seen growing on the Santa Ynez Reservation, but that *saus chino (Salix laevigata)* would *not* be used for that purpose (Robert Ontiveros, personal communication 1978). This may indicate that the Chumash did not regard red willow as useful for medicinal purposes.

ARROYO WILLOW
Salix lasiolepis

Certain utilitarian items reportedly employed this species. Stoppers of arroyo-willow leaves or twigs were inserted to plug the mouths of basketry water bottles. María Solares had seen fishing poles made from *shtayɨt*, and Mary Yee said it was good for making switches or whips. Fernando Librado said that only willow was burned in the sweathouse; María Solares noted that *shtayɨt* firewood was more easily chopped than *wak (S. laevigata)*, which had a twisted grain.

Lamiaceae — Mint Family

Salvia spp. — **SAGE** (general)

(no generic Chumash name known) — (Sp) Salvia

In local Spanish, the general category "*salvia*" refers to woody perennial shrubs only: white sage (*Salvia apiana*), purple sage (*S. leucophylla*), and black sage (*S. mellifera*). No corresponding general term for "sage" was recorded in Chumash languages. They considered the annual chia sage (*S. columbariae*) to belong to a separate category, which may also have included thistle sage (*S. carduacea*) and hummingbird sage (*S. spathacea*). Chumash speakers never applied the Spanish term "*salvia*" to any of these three herbaceous species.

Where possible, information on Chumash use of particular species of sage is listed separately below. There is no clear indication that Chumash people made any use of *Salvia leucophylla* or *S. mellifera*.

Historical sources mention medicinal uses of unidentified species of sage by the Chumash and by non-Indian settlers in the area. *Salvia* leaves were simmered and the decoction drunk at bedtime as a remedy for night sweats (Birabent n.d.). In the northern Chumash region, *salvia* was boiled in milk and taken for insomnia, made into a hot tea to induce perspiration, and used as a wash for swelling (Garriga 1978:32, 34, 40). Early settlers took a fluid extract of *salvia* for stomach ailments (Benefield 1951:22, 25).

In the 1970s, Indian and Hispanic people in the Santa Barbara area used tea of "many varieties" of sage leaves as a general medicinal tonic to cleanse and strengthen the body, especially the blood and nervous system. Sage tea was taken for anemia, indigestion, colds, and flu (Weyrauch 1982:18)

Lamiaceae — Mint Family

Salvia apiana JEPSON — **WHITE SAGE**

(B, I, V) khapshɨkh — (Sp) Salvia Mayor, Salvia Real

The Spanish names for white sage are descriptive of its larger leaf size and overall height compared with other sages. The Chumash sometimes called it by the descriptive term *'alikhshanuch* (V), meaning "ashy," referring to the color of the leaves.

Unlike some other California peoples, the Chumash did not eat the seeds of white sage, but they did peel and eat the young, tender stem

WHITE SAGE
Salvia apiana

tips. Fernando Librado said that a hunter placed the leaves of white sage in his mouth so that the deer would not detect his presence. Candelaria Valenzuela indicated that acorn granaries were lined with leaves of this plant. Perhaps its aromatic qualities were thought to repel insects and other pests.

María Solares recommended placing the fresh, strong-scented leaves of white sage on one's head as an effective treatment for headache. Luisa Ygnacio said the pounded leaves could be added to cold water and drunk to induce vomiting. The Chumash considered it beneficial to purge the body in this way. More recently, it has been reported that the plant was used for the nervous system and "to subdue too strong a sex drive" (Weyrauch 1982:18). This latter use has appeared widely in popular literature, almost always stated in these exact words, which suggests that it has been passed on from a single source and does not reflect traditional Chumash usage.

For Chumash people today, white sage is most important for burning as a purifying incense at ceremonies and gatherings. The young branch tips are dried slowly, pressing the leaves with the fingers periodically to make them lie flat on the stem. Sometimes yarn or string is wrapped around the sprig to facilitate shaping the sage bundle. After it has been thoroughly dried, the tip of the bundle can be set afire. It is allowed to burn for a few moments and then the flame is blown out, so that the leaves continue to smolder and release a thick, aromatic smoke. Inhaling the smoke and allowing it to waft over the body are thought to promote spiritual balance and harmony. These sage bundles are also presented as offerings and as gifts. Ubiquitous as it is today, use of white sage in this manner is not mentioned in Harrington's notes from interviews with earlier Chumash people. It is not certain how this practice originated among the Chumash, though it may have developed from similar uses of California sagebrush (see *Artemisia californica*).

Lamiaceae — Mint Family

Salvia carduacea BENTH. — **THISTLE SAGE**

(I, V) pakh — (Sp) Chia Gruesa

Harrington's notes indicate that the Chumash recognized two kinds of chia, one coarser (*Salvia carduacea*, thistle sage) and the other finer (*S. columbariae*, or "true" chia). The Spanish descriptive term acknowledges the similarity of the two species and observes that thistle sage

THISTLE SAGE
Salvia carduacea

has coarser leaves and larger seeds. Though the Chumash sometimes ate the seeds of thistle sage, they vastly preferred those of true chia.

Harrington's Kitanemuk consultants had a similar view of the two species. They referred to a specimen of thistle sage as "another kind of chia" and said it was not very good tasting, but they ate it when true chia was scarce (Timbrook 1986:53).

Lamiaceae — Mint Family

Salvia columbariae BENTH. — **CHIA**

(B) 'ilépesh — (Sp) Chia

(I) 'i'lepesh

(O) l'ɨpɨ

(V) 'itepesh

Chia was one of the favorite foods of the Chumash and many other California peoples. The term is Nahuatl (Aztec) in origin. It was adopted into Spanish to refer to certain plants in the genus *Salvia* that are cultivated in Mexico, and the small, oily seeds they produce. Among Chumash peoples, both the word "chia" and the native names given above refer to the plant and seeds of only one species, *Salvia columbariae*. A "coarser kind of chia" (thistle sage, *S. carduacea*) was recognized by some consultants but was much less prized. Both these species are herbaceous annuals that flower in the spring, produce seeds, and die. Broader application of the term "chia" to include seeds of woody perennial sages, sometimes seen in the literature, does not reflect Chumash classification or usage (see Timbrook 1986).

Early historical accounts indicate that chia seeds were gathered in some quantity. Harrington's consultants noted that in the early twentieth century, the plants were no longer seen in places where they were formerly common. The decline in abundance of chia may be due in part to habitat changes resulting from the introduction of new plant and animal species by European colonists, and from the suppression of Chumash grassland burning practices in the late eighteenth century (Timbrook, Johnson, and Earle 1982).

Even when chia was plentiful, the volume of the harvest was certainly not comparable to that of acorns (*Quercus* spp.) or of *islay (Prunus ilicifolia)*. But it was highly valued and nutritious, and perhaps more important in the qualitative sense than in terms of the quantity actually used (Timbrook 1986).

Chumash people collected chia seeds in late spring and summer with the aid of a seedbeater—an openwork basket with a handle, much like a tennis racket in appearance, woven of willow (*Salix*) or sumac (*Rhus trilobata*). Some women employed a hooked stick to pull the seed heads over the shallow gathering basket before striking them with the seedbeater. The gathering basket was a foot in diameter and six inches deep, with a flat bottom. It was placed on the ground or held between the knees so the seeds fell into it. Each time it got to be half full, the woman emptied it into the burden basket. When she had finished gathering she carried the burden basket back to the village in a net on her back using a tumpline.

Another harvesting method was to pull up the whole plants, pile them up, and thrash them with the seedbeater to knock the seeds out. The seeds could then be swept up and winnowed to remove dirt and other foreign matter. In either case, many seeds escaped the harvest, so that new plants would grow to provide a seed crop the following year.

Chia—like acorns, *islay*, and red maids (*Calandrinia*)—was stored in people's houses in large baskets. These were about two feet across and two and a half feet high, woven of *Juncus* rush and placed on tule (*Scirpus*) mats supported by sticks to keep out moisture and pests. Chumash people continued to store foodstuffs in the same way in mission times, placing the baskets of chia and other dried seeds on six- or eight-foot-high shelves in their adobe houses.

To prepare chia for eating, the Chumash usually toasted the dried seeds and then ground them to a fine flour. Toasting was often done by tossing the seeds in a basketry tray with hot coals made from live-oak bark. A rapid, continuous motion prevented the coals from scorching the basket or melting the thick layer of tar with which it was generally coated. Many Chumash preferred to toast chia seeds in a steatite *olla* placed directly on the fire; with this method the seeds were less likely to burn. A fragment of broken *olla* could also be used. The toasted chia seeds were called *molɨsh*. The next step was to grind them with a stone mortar and pestle.

Chia was never made into a boiled mush as acorns or *islay* were. Instead, after being toasted and ground, it was usually mixed with cold water and stirred for a few minutes until it thickened. One could vary the consistency of this drink, called chia pinole, according to individual taste. When mixed thin, it improved the taste of drinking water and was a good thirst-quencher (Rothrock 1878:48). Some people liked to eat the thickest part of the gruel and then add more toasted chia meal to the remaining broth to make an even richer dish. Or, the thicker mixture could be allowed to dry into cakes or loaves (Craig 1967:126). Neither the chia drink nor the loaves were cooked further. Chia was mixed in a wooden bowl (Henshaw 1955:153).

Several variations on the chia pinole theme were reported. Fernando Librado said that a nice drink could be made from chia and oats (*Avena*) toasted and ground separately, then mixed together into cold water with a little salt. Oats were introduced to the Chumash by the Spanish; so was corn, which some people added to chia to improve the taste. Perhaps seeds of native grasses were once used in the same way. According to Candelaria Valenzuela, "half-breeds" sometimes added sugar to the chia pinole (Blackburn 1963:147). A teaspoonful of whole raw or toasted chia seeds could be put into a glass of cold water, with or without sugar, to make a drink which was said to be especially refreshing and fine for the stomach in hot weather. Fernando never saw Indian people drink raw chia in water and implied that it might be a custom of the Spanish.

The chia gruel was taken between bites of other foods, not by itself. Foods that were eaten along with chia included meat, abalone, molded cakes made from juniper berries or other meal, and toyon berries (*Heteromeles*). The Chumash said that chia was particularly good with *carrizo* sugar (see *Phragmites*), which tended to stick to the teeth. The consumption of chia pinole and cakes declined with the loss of native culture and language, and these foods were no longer valued by modern descendants in the 1950s (Gardner 1965:285), although some mix chia seed with water and sugar in the Mexican fashion to make a beverage (Weyrauch 1982:18).

Probably the most widely published phrase in California ethnobotany is Bard's assertion that "one tablespoon of these [chia] seeds was sufficient to sustain for 24 hours an Indian on a forced march" (1894:4). Many subsequent authors have repeated this statement. Some, in dramatic exaggeration, reduced the amount required to a teaspoonful (Saunders 1914:134–136; Balls 1962:25). Though chia is a nutritious food, these claims seem extreme.

Proximate analysis of chia seeds shows them to be rich in fat and protein, 20 percent and 22 percent respectively. About half a cupful, 100 grams, yields 450 calories (Gilliland 1985:46). Thus a tablespoonful (12.5 g) would provide 56 calories, 2.75 grams of protein and 2.5 grams of fat. A teaspoonful would only offer a third as much. Though

CHIA
Salvia columbariae

perhaps fairly filling when beaten into cold water, either amount would be scant rations for twenty-four hours' work. Chia does have one and a half times as much protein and twice as much fat as other sage seeds available to the Chumash. Good-tasting and satisfying, chia was highly regarded among every Native American group in whose territory it occurs (Timbrook 1986:53–54).

Because it was such a favored food, chia was also a frequently traded commodity. Quantities of the seeds were measured in women's basketry hats as a unit of exchange. According to Fernando Librado, one hatful of chia was worth five hatfuls of acorns, and a man who wanted a good elderberry-wood bow would give five to eight hatfuls of chia for it. Although the chia plants grew on Santa Cruz and Santa Rosa Islands, it was common for the island Chumash people to come to the mainland to buy chia seeds rather than relying exclusively on their own supplies. In exchange for as many hatfuls of chia, acorns, *islay*, and red maids seeds as they could carry, María Solares said, the islanders gave strings of olivella shell money they had made.

As a medicinal substance, chia is most famous as a material to clear the eyes of irritating particles. To do this, both Chumash and Spanish people would put one or more chia seeds next to the eye, under the lid, often before going to bed; Fernando Librado said it removed the heat from the eyes at night. As moisture from the eye caused the seeds to become soft and glutinous, foreign bodies adhered to the seeds and could then be removed (Bard 1930:23; Gardner 1965:298; Weyrauch 1982:18). Simplicio Pico told Harrington that José Rios was able to remove a bit of iron from his eye by this method, "like mercury gets gold in mining," while waiting for the doctor to arrive. The seeds were also useful as a poultice for wounds. Chia gruel, though taken as food, was said to be a beneficial treatment for inflammation of the stomach and bowels, dysentery, and hemorrhoids (Rothrock 1876:212).

Chia seeds were given as a ceremonial offering at most public and private rituals. At any time of year, individuals left chia and other seeds, bead money, and tobacco (*Nicotiana*) at shrines. During the fall harvest festival, the dancers tossed chia and other seeds over the crowd. At the winter solstice ceremony itself and at the gatherings held in preparation for the ceremony, chia was given as an offering to the Sun and its representatives (Hudson et al. 1977). At various gatherings, Chumash people threw chia seeds into the fire as an offering. For example, after a funeral, a ritual leader built a fire in the center of the house where the body had been lying. As he sang, he threw green *romerillo (Artemisia californica)* branches and dry chia seeds into the fire. The oily seeds flared up in the fire, in dramatic punctuation to prayers and speeches on this and other solemn occasions (see Hudson, Timbrook, and Rempe 1978:163–167).

Charmstones, believed to protect homes from wind and fires, were kept in a box with chia seeds, tobacco, and money. At the end of each year this "food" was ceremonially burned and replenished. Certain types of charmstones were used in a ritual to bring benefits to the community—to make rain, put out fires in the mountains, call fish up the streams, and the like. Chia and white goose down were sprinkled over the stones, to the accompaniment of rattles and chants (Henshaw 1885:110–113; Timbrook 2000).

Important as chia was in Chumash ceremonial practice, it may not have been used quite as extensively as once believed. Large quantities of small black seeds found with burials in the Chumash region, formerly thought to be chia, have since been identified as red maids, *Calandrinia ciliata* (Timbrook 1986:58–59). Nonetheless, many California peoples certainly did gather large amounts of chia. One Nomlaki woman reportedly had "a remnant of six or seven pounds" of *Salvia columbariae* seeds in her possession, having gathered them in the Sacramento Valley the previous year (Chesnut 1902:384). The Cahuilla periodically burned stands of chia to improve production (Bean and Saubel 1972:136).

Lamiaceae — Mint Family

Salvia spathacea E. GREENE — **HUMMINGBIRD SAGE**

(B) qimsh — (Sp) Diosa, Diosita

(V) pakh

Lucrecia García collected three specimens of *Salvia spathacea* about 1928 and labeled them with Spanish and Barbareño names: *diosita* and *kimsh*, respectively. Fernando Librado said that another Spanish name for *diosa* was "*borraja silvestre*," which he called *pakh* in Ventureño. A decoction of these leaves was drunk or used as a bath in the treatment of pulmonary ailments and rheumatism. Fernando Librado described the use of *diosa* as part of a treatment to cure an illness caused by sorcery: fresh, juicy *diosa* leaves were rubbed all over the patient's body, and he lay on a bed of *diosa* leaves and also drank a tea made from them. The sorcerer who had caused the illness died when he learned the patient had recovered. The full account was published by Hudson (1979:59–60).

HUMMINGBIRD SAGE
Salvia spathacea

Caprifoliaceae | Honeysuckle Family

Sambucus mexicana C. PRESL. | **BLUE ELDERBERRY**

(B, Cr, I, V) qayas | (Sp) Saúco

Some sources refer to the wood of the elderberry tree as "elder," thereby leading to potential confusion with alder (*Alnus rhombifolia*) or box elder (*Acer negundo* var. *californicum*). The Chumash clearly employed *Sambucus*, not these other species, for the uses described here. The plant produces edible berries, but only one of Harrington's consultants mentioned eating them: Simplicio Pico said that *saúco* fruit was used for pies. Elderberry was far more important as a source of wood for tools and musical instruments, and as a medicinal plant.

The Chumash favored elderberry wood for self bows used in hunting small game. Since these bows were not backed with sinew, they were not spoiled by sea water, so mainland Indians often took them to the islands to hunt sea otters. They also sometimes hunted larger game, such as deer, with elderberry bows. Wearing a deer decoy headdress, the hunter was able to get within ten feet of the animal before shooting. This was necessary because unbacked bows had a much shorter range despite their greater length.

According to Fernando Librado, the Chumash bow makers' guild "looked after both kinds of bows"—that is, they regulated the manufacture and possibly the ownership of both sinew-backed and self bows. To make a self bow, the bow maker selected a section of elderberry branch about an inch and a half in diameter and four feet long, straight and free of knots. He cut the stave from the green tree. If it split while drying, said Simplicio Pico, that meant the wood had been cut at a time when it contained too much sap, such as at the full moon.

Elderberry shoots and branches are filled with a soft pith that can easily be removed to leave a hollow tube useful for a variety of purposes. By plugging the ends of these tubes, Chumash people made small tobacco containers which were worn tied to the belt at the left hip. Though usually of stone, smoking pipes were sometimes made of wood, and elderberry was preferred because it was naturally hollow and required less working (Craig 1967:124). Dance wands had handles of elderberry with a feathered foreshaft of willow inserted into the hollow center. The feathered pole erected at the winter solstice might also be made from a thick piece of elderberry wood.

Chumash people made several kinds of musical instruments from *Sambucus* stems. These included flutes, which they played for enjoyment or in courtship. The flute was held at a diagonal angle, blown from one end, and played with the first two fingers of each hand. The slightest flaw in an elderberry shoot made it useless for a flute. It had to be about two feet long and an inch and a quarter in diameter, and the joints in the stem could not be too close together. The shoot was roasted in a pile of embers with sandy dirt placed at each end to reduce respiration. After "sweating" the stick in this way, the flute maker extracted the interior pith with a reaming rod of toyon (*Heteromeles*) that had been straightened with a grooved stone. He inserted the rod into the elder stem repeatedly, twisting and pulling it out frequently. Then he dried the shoot for two or three days before removing the bark. He would typically drill four, or possibly six, fingering holes into it and paint the flute or decorate it with inlay of *carrizo* and shell.

During the Spanish period, Franciscan missionaries at San Fernando, San Buenaventura, and San Luis Obispo mentioned local Chumash people playing flutes made of elderberry (Geiger and Meighan 1976:133–134). After the missionaries introduced the giant reed (*Arundo donax*), the Chumash began making flutes of that instead.

Chumash musicians often accompanied singing and dancing, including the Bear Dance and many others, with a split-stick rattle, or clapperstick, known as *wansak* in Barbareño and *'akskatata* in Ventureño (Hudson et al. 1977:81–83). To make this rattle, they would select a straight, eighteen-inch-long piece of elderberry or *carrizo* and split it for most of its length, leaving an unsplit section at one end for the handle. It was sometimes necessary to carve away a bit of wood at the base of the split to make the rattle shake better, or to wrap the handle to prevent the stick from splitting further. The clapperstick is played in several ways: by striking it against the hand or another part of the body, by bringing it to a sudden pause in the air, or by rapidly shaking it to produce rhythmic clapping sounds.

In some dances, bullroarers served as musical accompaniment. The bullroarer was a flat stick of elderberry or willow, nine inches long, with a nettle (*Urtica*) fiber string attached to one end. It was played by whirling it around overhead. When sacred activities were taking place in the ceremonial enclosure, an old man would use the bullroarer, with its distinctive whirring sound, to warn away the uninitiated.

Fernando Librado also described musical bows made of elderberry, but it is possible these had been introduced from Mexico in historic times. He said that this instrument was an elder stick nearly two inches thick with a foot-long section hollowed out on the belly side. The string was made of sinew, gut, or Indian-hemp (*Apocynum*) fiber and tightened with a peg at one end. The opposite end of the bow was held in the mouth, which acted as a resonating cavity, or the bow could be placed on a box to augment the sound.

Elderberry was one of the woods that the Chumash used in making firesticks, either just the vertical drill or both the drill and the horizontal hearth piece. The drill was inserted into a pit in the hearth stick and rotated rapidly between the hands to produce glowing hot sawdust, which could be blown into a flame. Normally, however, people preferred to keep coals of oak bark buried overnight and use them to rekindle a fire rather than start all over again with the firesticks.

Other minor uses of elderberry wood have been reported. The pump drill, thought to have been introduced to the Chumash in historic times, had a shaft of elder with a flint point attached. This tool eased the task of perforating beads and ornaments. Chumash men and boys at Ventura used a two-foot rock-throwing stick of elder or *Arundo* in sham fights. With these sticks, they could throw stones accurately enough to hit a target at forty paces. Finally, Juan Justo and Fernando Librado mentioned the use of woven bags made of *saúco* bark for collecting acorns or *islay;* however, it is possible that they meant *saus* (willow, *Salix*).

Different parts of the elderberry plant had a number of medicinal uses. Indians and settlers drank a decoction of the flowers to treat colds and fever (Birabent n.d.; Blochman 1894b:40; Benefield 1951:22, 26). Some also used the leaves to treat colds and fevers (Bard 1894:6–7). The hollow elderberry stems may have been used in medical treatments; a nineteenth-century physician reported that the Chumash made syringes by attaching an elder tube or bird bone to a bladder. He also wrote that they performed bloodletting by inserting an elder or bone tube into an incision and sucking (Bard 1894:8, 9). Some authors believe that the syringe at least was introduced to the Chumash in historic times (Walker and Hudson 1993:50). Other treatments in use at the missions included a poultice of elderberry buds on the head for sunstroke, and incense of elderberry flowers or ashes from the young branches as a treatment for wounds (Garriga 1978:23, 42–43).

Harrington's Chumash consultants said that elderberry flowers were gathered and dried in the sun for later use. To induce sweating, which was part of many medical treatments, María Solares recommended that patients drink a cup of water in which elderberry flowers had been boiled, and soak their feet up to the knees in hot water to which ashes had been added. Elderberries mixed with egg white and mud were applied as a plaster to an aching part of the body. According to Luisa Ygnacio, people would pound the root of *saúco*, boil it, and drink this hot as a very strong laxative tea. To counteract the powerful effect of this treatment, peppergrass (*Lepidium*) seeds were pounded and drunk in cold water. Fernando Librado had been told of splints made from strips of elder wood, used to hold poultices over wounds.

Santa Ynez Chumash interviewed in 1961 said they soaked injured parts of the body in a decoction of boiled elderberry flowers (Gardner 1965:298). More recently, Antonio Romero told about his mother using these flowers to close wounds. Once, when she got a severe cut from broken glass, she boiled elderberry flowers and applied the tea as hot as she could stand it; the cut closed up right away (Antonio Romero, personal communication 1978). In the Santa Barbara area about 1978, Indian and Hispanic people reportedly made fresh elderberry flowers into a tea drunk as a blood purifier; to induce sweating and thereby shorten the course of measles and fever; and to treat sore throats, colds, and constipation (Weyrauch 1982:19).

Fernando Librado said that a slender tube of elderberry wood could be used in love magic. The heart or pith was drilled out with a stick of chamise (*Adenostoma*) or ceanothus. Into this tube a woman put a little dirt from where the man she desired had spit, dirt from where he had urinated, dirt from his clothes or skin, and hair from his temple. She placed the tube on her doorsill, and probably said prayers as well. In fifteen days the man would come, looking around like crazy. Men used the same method to get a woman they wanted.

This multipurpose plant also offered protection from rattlesnakes. Luisa Ygnacio was advised to mix elderberry leaves with human urine and rub it on her legs before going into brushy places to cut wood. This precaution would supposedly ensure that the snake would rattle a warning before she got too close.

Lamiaceae — Mint Family

Satureja douglasii (BENTH.) BRICQ. — **YERBA BUENA**

Mentha spp. — **MINT**

(V) 'alaqtaha — (Sp) Yerba Buena

Yerba buena is usually identified as *Satureja douglasii*, but no voucher specimens of *yerba buena* are extant in the Harrington collection to confirm its attribution among the Chumash. Although it is found in moist areas near La Purísima and elsewhere in northern Chumash territory, *Satureja* is not common on the coastal side of the Santa Ynez Mountains.

There are indications that when Chumash people refer to "*yerba buena*" they are thinking of mint. Ventureño elders given sprigs of *Satureja* agreed among themselves that it was "*yerba buena del campo*"—*yerba buena* of the countryside—and indicated that this was not the kind they usually used (Vincent Tumamait and Bertha Blanco, personal communication 1987). One species of mint, *Mentha arvensis*, has a broad native distribution but at least some plants seen here are naturalized from Europe and not native to this region (Munz 1959:717; Hickman 1993:716). Garden mints are widely cultivated and sometimes escape into the wild. Medicinal use of either type of *yerba buena* may have been introduced to the Chumash, though it has been common since the late nineteenth century.

Indians and early settlers in the Ventura area thought that *yerba buena*, *Micromeria* sp. (a former name for *Satureja douglasii*), was an effective treatment to destroy parasitic worms, expel gas to relieve colic, promote menstrual flow, and reduce fever (Bard 1894:6). An early source in Santa Barbara said that *Micromeria douglasii*, the well-known *yerba buena*, was found in moist soil in the woods and was valued as a blood purifier (Bingham 1890:36). Another author noted that *yerba buena*, wild mint, was used for inflammation of the bladder and stomachache (Birabent n.d.). A missionary in the Obispeño area recommended tea of *yerba buena*, or mint, against worms and diarrhea (Garriga 1978:25, 42).

Harrington's consultants, however, barely mentioned *yerba buena*. Simplicio Pico said that to treat eyes sore from a cold, "*yerba buena silvestre*, what you call in English wild peppermint," was boiled in water and the patient leaned over it to let the aromatic steam into the eyes. In Ventureño Chumash this plant was called *'alaqtaha*, meaning "both warm and cordial." Pico added that one should also stay in the house out of light and wind for fifteen days to complete the treatment. Another consultant described covering a butchered beef head with *yerba buena* when roasting it in the earth oven.

In the late 1950s, Santa Ynez Chumash said *yerba buena* (not identified from a specimen) was a popular tea for the stomach and as a beverage, and that meat was formerly cooked with this plant in earth ovens (Gardner 1965:298). In the late 1970s, *yerba buena* (mint) was the first herb tea mentioned by most Chumash and Hispanic people and one of the most often used. Either fresh or dried leaves were placed in boiling water and the tea drunk for relaxation, to soothe the stomach and reduce gas and colic in babies. The fresh leaves could also be rubbed on sore muscles (Weyrauch 1982:20–21).

Medicinal use of *yerba buena* was reported among other California Indians. Harrington found that the Ohlone took it for toothache and pinworms, but he did not preserve a specimen of the plant for botanical identification (Bocek 1984:253). The Cahuilla drank *yerba buena (Satureja)* tea to reduce fevers and cure colds, and they applied the plant to the head for headache (Bean and Saubel 1972:139). *Yerba buena* has been widely used in medicine among Hispanic and some Indian peoples throughout the Southwest and northern Mexico, where the term most often refers to *Mentha spicata*, the garden spearmint (Ford 1975:327–330).

Schoenoplectus spp.: See *Scirpus*

Cyperaceae — Sedge Family

Scirpus spp. — **TULE, BULRUSH** (general)

(no generic Chumash name known) — (Sp) Tule

It is uncertain whether a general term equivalent to "tule" existed in Chumash languages. The Chumash recognized three main types of tule, which Harrington used the Jepson (1925) manual to identify as follows: "round" (*Scirpus acutus* and *S. californicus* were lumped into a single folk species), "triangular" (*S. americanus* and *S. pungens*), and "wide tule," or cattail (*Typha* spp.). Those are among species in the genus *Scirpus* that have been reassigned to *Schoenoplectus* (Flora of North America Committee 1993+). In this book I continue to use the earlier nomenclature, which is easier for most readers to look up in the common reference works.

This section covers miscellaneous uses for which no particular type of tule was specified, as well as some of the purposes for which more than one kind could be used. Subsequent sections for each named type of tule describe the uses of that particular species.

Tule was the most common material that Chumash people employed for thatching houses. Any kind of tule could be used (see also *S. americanus*, *S. californicus*, and *Typha* spp.). The stems were laid up against the framework of the house in overlapping layers like shingles, starting at the bottom. Two men, one on the inside of the house and one on the outside, worked to tie the thatching onto the framework. They used a special wooden needle made of toyon (*Heteromeles*) or scrub oak (*Quercus berberidifolia*), half an inch wide and two feet long, with an eye near the point. With this needle, they threaded ties made from cordage or soaked rawhide thongs through the thatch, over the crosspiece of the house framework, and back out through the thatch. They added more courses of thatch as needed to cover the house, leaving an opening at the top to serve as a smokehole. Most houses had doors made of woven tule. The house of a chief was lined with tule mats placed under the thatching to keep the thatching from falling through or looking unsightly.

Inside, the house was divided into rooms by hanging tule mats as partitions. People also slept on mats, either on raised platforms or directly on the floor, and sat on them when doing many kinds of work. Mats had a number of other uses: enclosing the ceremonial structure, or *siliyɨk;* making windbreaks; shading plank canoes; storing feathers and other regalia; and wrapping the dead for burial.

The Chumash made two basic kinds of mats: twined (usually of triangular tule, *S. americanus/S. pungens*) and pierced (round tule, *S. acutus/ S. californicus*). Additional information is provided under these species, below. Ordinarily, mats were the length of the tule stems used, but longer ones could be made by using two tule stems placed in opposite directions with the tips overlapping. The ideal length for burial mats was eight feet. Mats could be made in different colors, Fernando Librado said, depending on how the tule was cured.

During canoe construction, the vessel had to be shaded with tule mats to allow the tar adhesive to harden before adding another round of planks. After the canoe was finished and ready for use, it still needed to be kept in the shade. The old Ventura canoe builders stored their canoes in the tule marsh at the mouth of the Ventura River. They cut and piled up tule stems so the canoe could rest out of the water, and bent the tule growing on both sides over the canoe as a sunshade. The tips of the tule stalks interlaced like the fingers of clasped hands, and a pole was laid on top to hold them in that position. This prevented the sun from drying out the wood of the canoe and softening the tar caulking. Canoe construction procedures and use are described in Hudson, Timbrook, and Rempe (1978). The Chumash made another sort of watercraft, the balsa canoe, from tule stems; see *Scirpus californicus.*

The mouth of a tarred basketry water bottle was sometimes stoppered with a bunch of tule stems, according to Luisa Ygnacio. Otherwise, Harrington's Chumash consultants denied that tule was used in basketmaking (1942:23). A comprehensive study did find a few examples of Chumash coiled baskets which had sewing strands of split bulrush root (*Scirpus* sp.), presumably dyed black by burial in mud (Dawson and Deetz 1965:202). These are relatively rare; perhaps they were made in a part of Chumash territory not studied by Harrington, or at an earlier date. The Chumash also made twined basketry of tule (see *S. californicus*).

The framework for the Fox Dance headdress was of three-leaved sumac (*Rhus trilobata*) covered with woven tule. The long braid that hung from the headdress was made of two tule hearts and a two-ply milkweed-fiber (*Asclepias*) cord; willow bark or rags could also be used to make this "tail."

Chumash archers tied dry tule stems together into firm bundles for use as targets. One type, about three inches thick and five or six inches long, was mounted on a slender stick. Men and boys shot at this target with arrows that had hardwood foreshafts. Another archery target was much larger, about four feet long, and the archers stood about twenty paces away to shoot. If an arrow went half its length into one of these targets, it was considered good for hunting big game.

Ritual and medicinal practices also made use of tule. An unusual method of ingesting tobacco (*Nicotiana*) was practiced at Santa Ynez Mission in gatherings of men who were knowledgeable in traditional Indian religion, according to María Solares. They dipped a tassel made of the shredded, sponge-like pith of tule stem into a mixture of liquid tobacco and sucked it. Henshaw described another ritual use of this plant: at the time of initial menstruation, girls were required to remain for five days in a pit preheated with hot stones and lined with tule (Henshaw 1955:158). Ashes of burned tule were reportedly massaged onto the skin as a treatment for rheumatism, and broken bones were splinted with twined tule stems coated with asphaltum (Bard 1894:8–9).

See *Scirpus acutus/S. californicus* for additional medicinal uses of tule.

Cyperaceae | Sedge Family

Scirpus acutus MUHL. EX BIGELOW var. *occidentalis* (S.WATSON) BEETLE

BULRUSH, TULE

(Sp) Tule, Tule Redondo

(B, Cr) stapan

(I) swow

(V) kawɨyɨsh

Scirpus californicus (C. A. MEY.) STEUD.

CALIFORNIA BULRUSH

(Sp) Tule, Tule Redondo

(B, Cr, I) stapan

(V) kawɨyɨsh

When speaking Spanish, Chumash people recognized three categories of tule: round, triangular, and wide (or cattail). They placed *Scirpus acutus* and *S. californicus* in the category of *tule redondo*, "round tule," even though stems of the latter species tend to be somewhat triangular toward the tip. These two species do hybridize with one another (Hickman 1993:1147).

In Chumashan languages, however, the picture is slightly different. Although Barbareño and Ventureño lump these two species into a single folk taxon, Ineseño assigns them separate names—*swow* for *S. acutus*, and *stapan* for *S. californicus*. These linguistic distinctions may reflect the distribution and relative abundance of the two species. Both are common in Ineseño territory where they were distinguished by separate names, but south of Point Conception, where *S. acutus* is relatively uncommon, it was included in the same folk category as the abundant *S. californicus*.

Another possibility is that the Chumash names *stapan* (B, I) and *kawɨyɨsh* (V) may have functioned as folk generic terms, much like "tule" in English and Spanish, as well as referring to particular kinds. Further linguistic research is needed to test this interpretation.

Justifiably famous as makers of plank canoes, the Chumash and their ancestors had also been making balsa canoes since ancient times. Quite distinct from the light tropical wood of the same name, the balsa is a type of watercraft made from bundles of tule tied together (see Cunningham 1989:35–40). Historically, the Chumash most often used balsa canoes on the calm waters of estuaries rather than on the ocean—Simplicio Pico had seen them in the lake at the Ventura River mouth, for example—although there are reports of island crossings made in balsas. According to Harrington's consultants, the Gabrielino and Yokuts as well as the Chumash used this type of vessel.

Some native California peoples are known to have made balsa canoes of *S. acutus* (Jepson 1925:153), and experiments with immersion of tule bundles indicate that round tule (identified as *S. acutus*, or possibly *S. californicus*) would be far superior for this use, since it retains its buoyancy even when repeatedly wetted (Richard Cunningham, personal communication 1983). I believe that this was the type used by Chumash boat builders. They cut green tules and allowed them to sun-dry for a couple of days. They put the craft together quickly, preparing three or five bundles that were stiffened with willow poles and bound together. They waterproofed the balsa by coating it with tar and then rubbing on fine powdered clay to remove the stickiness (Hudson, Timbrook, and Rempe 1978: 27–31).

Plank canoes also had components of tule. Even though the planks were glued and sewn together, the seams between the planks needed to be caulked after the hull had been completely assembled. The canoe builders separated the heart, or pith, from dry stems of tule (*S. californicus*), forced it into the seams with a bone tool, and coated the seams again with a tar mixture.

Tule baskets are the subject of conflicting information. According to Harrington's culture element checklist, all Chumash groups denied the use of tule in basketry and also denied using twined tule baskets. They did agree that their people used twined, asphalted, flat-bottomed water bottle baskets (Harrington 1942:22–23). Harrington's notes describe water bottles being made of *Juncus* rather than tule. In terms of actual specimens, however, a large number of twined water bottles found in dry caves and other interior sites are in fact made of tule (see Grant 1964:8–11). Tule water bottles are usually straight-sided, with well-defined shoulder and neck, while those made of *Juncus* are squattier, with sloping shoulders; both kinds have a concave bottom. The water bottle of tule may have been an inland type that Harrington's coastal consultants were not familiar with, or perhaps it had gone out of use before their day. Two-ply cordage of "*Scirpus lacustris*" was reportedly used as the warp in twining these water bottles, with the same material split and twisted in a single ply for the weft (Dawson and Deetz 1965:203). Probably the species used to make these bottles was actually *Scirpus californicus*, since *S. lacustris* is a European bulrush not found in California.

Most Chumash groups thatched their houses with tule (*S. californicus/ S. acutus*). These plants were in short supply on the islands, so other materials such as surf-grass (*Phyllospadix*), *carrizo (Leymus)*, or bracken fern (*Pteridium*) had to be substituted (Timbrook 1993:53). See the general information on *Scirpus* spp., above, for more details on thatching.

According to Harrington's consultants, the Chumash used *tule redondo (S. californicus)* to make mats of the pierced type only, not twined mats. They arranged the tule stems on the ground, placing the slender tip of one next to the thicker base of the next, alternating so that the sides of the mat would be parallel. The weaver stuck a slender wooden needle through about ten stems at a time and rotated it to hold the hole open. Through this opening she threaded cordage which was attached to the thick end of a *Juncus* stem. She held these pierced tules in place with her foot while she strung some more.

Pierced tule mats were called *tiyuqash* in Ineseño Chumash, *'as* in Barbareño. Mats with the tule stems drawn tightly together, called *'alaski* in Ventureño, were used for sleeping; looser ones, *pukash*, were used for roofing purposes. Inside the house, mats stood upright between beds to form separate sleeping quarters. Two tiers or courses of tule mats, making a height of six to eight feet, formed the walls of the *siliyɨk*, or sacred enclosure.

Cave caches in interior Chumash territory have yielded archaeological examples of mats. In one find, mats of both pierced and twined types had been used to wrap various artifacts. Pierced stems identified as *Scirpus californicus* and *Scirpus olneyi* (= triangular tule, *S. americanus*) were threaded with cordage of Indian-hemp (*Apocynum cannabinum*) at seven-centimeter intervals. Three types of twined mats were also identified as incorporating *S. acutus* and *S. californicus* (Grant 1964:6–7). Although Harrington's Chumash consultants said that these two species were only used for pierced mats, not twined ones, the archaeological finds suggest that actual standards were not so rigid, or that there was regional variation in mat-making technology. See also *S. americanus* and the general category of *Scirpus* spp.

Although most women wore skirts of animal hides, plant-fiber skirts worn by poor women could be made of *stapan* tule, or of the bark of cottonwood (*Populus*) or willow (*Salix*). A drawing in Harrington's notes of such a skirt shows hanging fringes with a few rows of twining at the top (Hudson and Blackburn 1985:27). Aprons or skirts of tules, tied in rows on strings of sinew or bark, were also described by Yates (1890:24; 1891:375).

In addition to carrying loads using a tumpline and net, Chumash people sometimes carried burdens on their heads, steadied and cushioned by carrying rings. These rings were simply made with stalks of round tule bent in a circle and wound around with more tule.

Food uses appear to have been relatively minor for these species. Luisa Ygnacio and María Solares said that the "root," or rhizome, of *stapan* (B, I) was eaten raw. Both these consultants knew that tule roots could be made into bread or dried and ground for mush, but they usually described these as food of the "Tulareños," or Yokuts.

María Solares knew of different medicinal uses for these two species. After the umbilical cord of a newborn infant was cut, the navel was sprinkled with ashes from the burned pith of "fat tule," *swow (S. acutus)* so that it would heal quickly. She said that dry ashes of *tule redondo*, or *stapan (S. californicus)*, were rubbed on the skin as a cure for poison-oak rash.

The plant name *swow* is preserved today in the modern town of Sisquoc, from Purisimeño *siswow*, "in the thick [-stemmed] tule" (Applegate 1975b:42).

Cyperaceae — Sedge Family

Scirpus americanus PERS. — **OLNEY'S THREE-SQUARE BULRUSH**

Scirpus pungens VAHL — **COMMON THREE SQUARE**

(B, I) swa' — (Sp) Tule, Tule Esquineado

(V) tup'

As noted above, the Chumash distinguished three types of tule: round-stemmed, triangular, and wide tule, or cattail (*Typha* spp.). Harrington consulted Jepson's 1925 manual and identified "*tule esquineado*," or triangular tule, as *Scirpus americanus*, even though some other bulrushes also have stems that are at least partly triangular. His consultants always referred to *tule esquineado* as "small" or "slender," perhaps to contrast it with the taller *tule redondo (S. acutus/S. californicus)* or with the slightly stouter triangular-stemmed bulrush species that have leafy stems (*Scirpus maritimus*, *S. microcarpus*, and *S. robustus*. The three latter species of bulrush do not seem to be included in the named category "triangular tule" and it is unclear whether they were used by the Chumash. Three square tule (*S. pungens*) is now recognized as a separate species but was formerly included in *S. americanus* (Hickman 1993:1148).

Triangular tule was used in thatching houses, but perhaps less commonly than round tule, cattail, or *carrizo (Leymus)*. Fernando Librado said that when people were lazy they just sewed loose stems of the *tup* kind of tule onto the house framework. More industrious people fastened two or three courses of woven mats onto the framework and covered it with *carrizo*. House thatching is also described under the other species of *Scirpus*.

Triangular tule was the only kind from which the Chumash made woven (actually twined) mats, according to Fernando Librado and María Solares. In this type, the tule stems were laid parallel and fastened together by twining across them with paired strands of either cordage or flexible tule stems. The ends and edges were braided. This type of mat was called *meshe'esh* (V) or *'untɨqɨsh* (I) and was used for sitting as well as for covering houses.

There are extant archaeological examples of this type of mat that are consistent in some ways with Harrington's data. These mats have rows of twining done either with the same material as the warp or with Indian-hemp (*Apocynum*) fiber cordage, and several specimens have edging of braided tule stems. Both triangular and round-stemmed types of tule, however, were used in these twined archaeological specimens (Grant 1964:6–7). The triangular species was identified as *Scirpus olneyi*—now *S. americanus*—and the round species as *S. acutus* and *S. californicus*. See the other species of *Scirpus* for more information on mats.

Chumash people used three-cornered tule to cover the large baskets in which they stored acorns and *islay* seeds. They placed the baskets on platforms and capped them with inverted baskets weighted with stones, then a layer of loose tule stems, and finally a woven tule mat held in place with a stick to keep the wind from blowing it off. The mat, the loose tule, and the basket lids were all removed each day so that the air could get in while the acorns were drying; but after about a week they were left on all the time.

Various kinds of special containers were fashioned from stems of slender triangular tule. Fernando Librado described a sort of oval basket about twenty-one inches wide, woven at one end and tied at the other, made to hold *panocha* (*carrizo* sugar or, later, brown sugar). It had a hinged oval cover, also woven of tule, and was lined with *carrizo (Phragmites)* leaves. Inside, the *panocha* was all in one big lump the shape of the basket. María Solares said these *panocha* containers were made from *swa'*. Another tule container, also of *swa'*, was used to prevent smoke and dampness from twisting or checking arrows stored indoors.

Tule containers for holding ceremonial paraphernalia included a type of bag, funnel-shaped (wider at the bottom) with a drawstring in which charmstones were kept, and a tiny mat in which a stone pipe could be wrapped when not in use. Feather dance regalia, including headdresses and skirts, were stored in tule receptacles, since a wooden box might injure the feathers, and removed every few days to air in the sun.

Cradles like little baskets, woven of *swa'*, were made for newborn babies, according to María Solares. For their first six months, newborns were covered with wildcat skins and placed in their cradles near the fireplace for warmth. After about six months they would get another type of cradle, which had a Y-shaped wood framework with a mattress and sunshade of *swa'* stems twined together with string. Candelaria Valenzuela made an example of this type of cradle for Harrington (Craig 1966:197; Craig 1967:93–94).

Place names incorporating the name of this kind of tule include *kaswa'*, which was near the mission asistencia at Cieneguitas, in Santa Barbara, and two locations in Ineseño territory (Applegate 1975b: 27, 33).

Taxodiaceae — Bald-Cypress Family

Sequoia sempervirens (D. DON) ENDL. — **COAST REDWOOD**

(B) wi'ma — (Sp) Pino Colorado

(I) wima' (see also *Pinus*)

(V) wima

> It was here, where the wi'ma does not grow and is hard to get, that the Indians made good board canoes. Wi'ma is driftwood, mostly redwood, which floats down the coast after a storm on the sea and comes from the west into the channel...Redwood was the best of the wi'ma from which to make a board canoe.
>
> —Fernando Librado
> (Hudson, Timbrook, and Rempe 1978:47)

The Chumash plank canoe, or *tomol*, is considered one of the greatest technological achievements of any North American Indian people. Redwood logs that had originated in central or northern California arrived on Chumash shores as driftwood already seasoned and ready for use. Canoe makers selected only the best logs, which were straight-grained, strong, and free from knots. Using bone wedges and hammerstones they split planks from these logs and fitted them together edge-to-edge. This process involved some bending of steamed boards and some shaping with stone, bone, and shell tools. The canoe makers glued each course of planks in place with a mixture of tar and pine pitch, allowing it to dry in the shade for several days. Next, they drilled holes near the edges of the planks and lashed them together with plant-fiber string. They built the canoe slowly, one row at a time. Finally they caulked the seams with tule pith and the tar-and-pine-pitch mixture. They painted the finished canoe with a mixture of red ochre and pine pitch as a sealant. Fernando Librado said that canoe paddle blades were also made of redwood, attached to a shaft of ironwood (*Lyonothamnus*). See Hudson, Timbrook, and Rempe 1978 for a complete description of canoe manufacture and use.

Redwood was one of the woods from which the Chumash made mortuary poles. Fernando Librado described these as two or three feet high and about five inches in diameter. They were painted with various designs, such as a chainlike pattern in red and white, so that the

THE TOMOL

The Chumash tomol, *or seagoing plank canoe, was a superbly engineered craft, built of split driftwood sewn together and caulked with natural tar. From the* tomol, *fishermen could lure large fish into harpoon range or catch smaller kinds with hooks or nets. The* tomol *was essential to the thriving trade between island and mainland peoples. These canoes plied the waters of the Santa Barbara Channel from A.D. 1000 until the mid-1800s.*

relatives of the deceased could recognize the grave. María Solares saw a grave pole at Tejon which was as tall as a telephone pole and six inches in diameter with the bark removed. She said these poles had to be made of *wima'*, so therefore they were very expensive and only certain people could afford them.

The Chumash used redwood for other purposes also. Pieces that were not suitable for canoe-making or other construction were used as firewood.

Santa Rosa Island was called *wi'ma* in mainland Chumash languages (Applegate 1975b:45). Fernando Librado told Harrington that the Chumash used to perform a Santa Rosa Island dance called *wi'ma*, the Redwood Dance, but he was unable to remember details (Hudson et al. 1977:83, 90).

Harrington noted that *wi'ma* was essentially synonymous with redwood, and the usages described here are confidently attributed to *Sequoia*. The Spanish term "*pino colorado*" may sometimes refer to at least one kind of pine and perhaps to other coniferous trees as well (see *Pinus*).

Caryophyllaceae — Pink Family

Silene laciniata CAV. subsp. *major* C. HITCHC. & MAGUIRE

FRINGED INDIAN PINK

(B) s'akhtutu 'iyukhnuts' — (Sp) Claverito, Clavelita

Lucrecia García labeled specimens of this plant with a Barbareño word meaning "hummingbird sucks it." This descriptive term was also applied to other plants with red, tubular flowers, such as bee plant (*Scrophularia californica*) and California fuchsia (*Epilobium canum*).

With another specimen of *Silene*, Harrington wrote a note with information recorded from Luisa Ygnacio in 1914. According to her, Marcelino *kwinayɨt*, who was a "captain," or chief, at Santa Ynez, said that women would boil this plant and drink the tea to make their menses flow, and they would take it with wine if they did not want to become pregnant. It was only this species, not any other with the same name, that was used in this way.

Iridaceae — Iris Family

Sisyrinchium bellum S. WATSON — **BLUE-EYED GRASS**

(B) sh'ichkɨ 'i'waqaq — (Sp) Huitota

The descriptive name Lucrecia García gave to specimens of blue-eyed grass means, according to Harrington's accompanying note, "Frog's g-string." Although Chumash men did not usually wear clothing in pre-Spanish days, this terminology may indicate that loincloths were not purely a Mission-period introduction. No Chumash uses were recorded for this plant, but other California peoples employed it medicinally for a variety of purposes (Strike 1994:147; Mead 2003:398).

Solanaceae — Nightshade Family

Solanum douglasii DUNAL — **DOUGLAS NIGHTSHADE**

(B, I, P) 'aqulpop' — (Sp) Chichiquelite

(Cr, V) qolpo'op

Many members of the nightshade family contain dangerous glycoalkaloids and are best regarded as toxic. Although the solanine content decreases as the fruit ripens, color is not an adequate indication of ripeness. Douglas nightshade is less toxic than most, but is similar in appearance to other species from which cases of poisoning have been reported. Boiling is thought to destroy the toxic compounds found in the introduced weed *Solanum nigrum*, black nightshade (Muenscher 1975:208; Fuller and McClintock 1986:249).

Despite its potential toxicity, Fernando Librado, Juan Justo, and María Solares all said they ate the fruit of this plant and described the flavor as sour and bitter. The Chumash ate the berries both raw and boiled, and children often gathered them. In historic times, people cooked them with sugar and made them into pies.

Among the medicinal uses of nightshade, Simplicio Pico reported that he had been treated for poison-oak rash by crushing *chichiquelite* leaves between the hands and rubbing them on the affected area. Rosario Cooper recommended mixing the juice from *chichiquelite* leaves with salt for this purpose, and another of Harrington's consultants said the berries, mixed with salt, were a poison-oak remedy. Juan Justo said that Spanish people applied the leaves as a poultice for pain. Women mashed the young leaves in water, strained the liquid, and washed their hair with it to freshen the scalp, according to Fernando Librado.

Tattooing was not a common or widespread Chumash practice, but Luisa Ygnacio thought that a few individuals had some tattoos. Fernando Librado had seen Indian boys tattooing each other. They mashed nightshade leaves with a little water and painted the juice on their skin, then pricked the skin with a bundle of cactus thorns tied together. Some people used sewing needles to make the designs. Charcoal from yucca stalks may have been mixed with juice from nightshade leaves or rubbed onto the tattoo separately for the color.

The Cahuilla used the fruit of Douglas nightshade to treat sore eyes and possibly as a dye (Bean and Saubel 1972:140). In the Southwest, leaves of *chichiquelite (Solanum nodiflorum*, now *S. americanum*) were taken internally to induce vomiting and expel worms, and applied as a poultice for pneumonia, rheumatism, and sores (Ford 1975: 179–180, 376, 424–425). The Nomlaki of northern California ate ripe fruit of this latter species, but the green berries have caused poisoning (Chesnut 1902:387).

Asteraceae — Sunflower Family

Solidago californica NUTT. — **CALIFORNIA GOLDENROD**

(B) stu 'imá' — (Sp) Oreja de Liebre

(I) shtu'ama'

(V) chtu 'ima

The Chumash names given by Harrington's consultants appear to be a translation of the widespread Spanish common name for this plant, referring to the resemblance of the leaves to jackrabbits' ears. Lucrecia García labeled three goldenrod specimens with another Chumash name: *smolush 'i'ashk'a*, "Coyote's mugwort" (see *Artemisia douglasiana*).

A small brush of goldenrod, snowberry (*Symphoricarpos*), or almost any other plant was employed to remove the tiny spines from prickly-pear cactus fruit (Hudson and Blackburn 1982:231). The principal Chumash use of goldenrod, however, was medicinal.

Simplicio Pico said that a decoction of goldenrod was taken for coughs. He also recommended it to treat sick horses, both as a drink and as a wash for bruises. Other early sources also mention the use of goldenrod for coughs and colds (Bard 1894:6) and as a wash and powder for sores or wounds (Birabent n.d.). These same uses were remembered by older Santa Ynez Chumash in the late 1950s: the ground leaves were put on wounds, and the decoction was used to wash wounds or drunk with sugar for cough (Gardner 1965:298). Santa Barbara urban Indians in the 1970s continued to take goldenrod for severe coughs and chest congestion, and they used it as a wash (though not as a powder or poultice) for healing wounds. The plant was also reportedly used as a general tonic, a hair rinse, a treatment for digestive troubles and liver ailments, and in feminine hygiene (Weyrauch 1982:10).

The Ohlone similarly treated sores, wounds, and burns with decoctions and powders of goldenrod leaves (Bocek 1984:255). The Cahuilla used it as a hair rinse and in feminine hygiene (Bean and Saubel 1972:140). Several other California peoples treated sores, wounds, skin irritations, toothaches, and sore throats with goldenrod (Strike 1994:148–149). *Solidago* was apparently not a widespread remedy in the greater Southwest, except as a poultice for sore throat in New Mexico (Ford 1975:237).

Caprifoliaceae — Honeysuckle Family

Symphoricarpos mollis NUTT. — **NUTTALL'S SNOWBERRY**

Lonicera subspicata HOOK. & ARN. var. *denudata* REHDER — **CHAPARRAL HONEYSUCKLE**

(B, I) tu' — (Sp) Oreja de Ratón, Escoba

(V) chtu 'iqonon

The Spanish and Ventureño Chumash names refer to the leaves, which were said to be the size and shape of a mouse's ear. The Barbareño, *tu'*, simply means "ear." One of the three *Symphoricarpos* specimens and two of *Lonicera* collected by Lucrecia García were labeled with this name. Simplicio Pico said *oreja de ratón* grows in willow thickets or woods in moist places, not on the dry mountainside. Luisa Ygnacio had seen much of it growing near Oak Park in Santa Barbara. Lucrecia's specimens were collected on a trip to Los Prietos along the Santa Ynez River.

Chumash people tied leafy sprigs or branches of this plant in bundles to make brooms and brushes for a variety of purposes. After prickly-pear cactus (*Opuntia*) fruits had been picked, the small spines would be removed with a brush of *oreja de ratón*, goldenrod (*Solidago*), or some other shrub (Hudson and Blackburn 1982:231). In mission times, people often attached a stick handle to larger bundles of the twigs for use in sweeping floors.

Other uses were also mentioned. Luisa Ygnacio said that *tu'*, or *escoba*, was used like *shuna'y (Rhus trilobata)* to make the coarsely twined utility baskets called *chiquihuites*. Boys used to make toy arrows from the stems of this plant, according to Juan Justo. María Solares commented that children sometimes ate the white, insipid fruits.

Brooms made from brushy stems of *Symphoricarpos rivularis* (now *S. albus*) were reported among the Ohlone (Bocek 1984:254).

Anacardiaceae — Sumac Family

Toxicodendron diversilobum (TORREY & A. GRAY) E. GREENE — **POISON-OAK**

(B, Cr, I, V) yasis — (Sp) Yedra

(O, P) wala

This is an appropriate place to remind the reader that I do not endorse any of the remedies discussed in this book. The Chumash regarded poison-oak as a useful medicinal plant, both applied externally and drunk as a tea.

Mission documents from the early nineteenth century describe plasters of *yedra* (a term that some authors later mistranslated as "ivy") as very effective in healing wounds. The priest at San Luis Obispo himself had seen a man who had been badly lacerated by a bear healed only with an application of powdered poison-oak (Geiger and Meighan 1976:75). About 1912, Chumash people were still familiar with the use of poison-oak juice to stanch the flow of blood from a cut.

At least one of Harrington's consultants considered the juice from poison-oak leaves and stems, freshly cut in early spring, to be the most effective remedy for warts, skin cancers, and other persistent sores. (The common Spanish term "*cáncer*" apparently referred to an ulcerated sore or canker, not what we call cancer in English.) If cauterizing the sore with *estafiate (Artemisia douglasiana)* was unsuccessful, said Fernando Librado, they would let it go until spring when the poison-oak sap was flowing well. They pulled the leaves from the stem one at a time and held them to the wound, allowing the juice—the "tears" of

POISON-OAK
Toxicodendron diversilobum

the poison-oak—to drip into it. Many leaves were required. The juice turned black as it touched the skin. When this black surface healed and fell off, the cancer was cured. For corns and calluses on the feet, Fernando used what he called "the Indian remedy"—cutting away the thickened skin and applying the juice oozing from a freshly cut poison-oak stem.

As a treatment for severe dysentery or diarrhea, the root of poison-oak was boiled, being careful not to allow the vapor to get into the eyes, lest blindness result. Consultants said that it was important to allow the decoction to cool thoroughly before drinking it.

Although many Anglo-Americans develop severe dermatitis from contact with poison-oak leaves and stems, this seems not to have been a problem for many Indian people. Indians may have had some degree of natural immunity to urushiol, the active component; perhaps they were also willing to tolerate a certain level of discomfort. The Pomo of northern California made baskets of poison-oak stems, wrapped food to be cooked in the leaves, and used the juice or ashes in tattooing (Chesnut 1902:364; Goodrich, Lawson, and Lawson 1980:82). The Ohlone made similar uses of the plant, but they reportedly did employ herbal remedies for poison-oak rash (Bocek 1984:250–251).

Harrington's Chumash consultants commented that different Indian groups had varying susceptibility to poison-oak. The Yokuts, a neighboring inland group, were said to be severely affected by it when they visited the coast, but the local Chumash were affected little or not at all. One woman said that when her mother had worked out in the field cutting poison-oak she got just a little rash on her arms.

Immunity seems to have diminished with the proportion of Chumash ancestry. Remedies for poison-oak rash, although known for some time, seem to have become much more important in recent decades. Harrington's consultants treated the rash by rubbing it with dry ashes from burned stems of tule (*Scirpus acutus*) or rush (*Juncus textilis*). Other remedies they mentioned were bathing the area in lime water—which the mission priests did—or mugwort tea (*Artemisia douglasiana*), or rubbing with leaves or berries of nightshade (*Solanum douglasii*) mixed with salt, or simply spitting on and rubbing the affected part if the rash was not severe.

By the late 1950s, Chumash descendants no longer made any medicinal use of poison-oak but eagerly sought remedies for its effects. Bathing in decoctions of mugwort leaves, coyote brush (*Baccharis pilularis*), or coffeeberry (*Rhamnus californica*) were among the remedies mentioned at that time (Gardner 1965:297–298). A decoction of coffeeberry leaves was also used as a poison-oak remedy by the Ohlone (Bocek 1984:250). Mugwort has now come to be regarded as a specific for poison-oak rash (Weyrauch 1982:9).

Some individuals claim that immunity can be obtained by spitting on the plant or by drinking a decoction of boiled poison-oak root (Weyrauch 1982:14; Romero 1954:11). Others who do not get the rash believe that their practice of eating a poison-oak leaf now and then has made them immune. None of these methods were reported by Harrington's consultants, although they did caution, as noted above, against the effects of the vapors that result from boiling the root.

Lamiaceae — Mint Family

Trichostema lanatum BENTH. — **WOOLLY BLUE CURLS**

Rosmarinus officinalis L. (Introduced) — **ROSEMARY**

(V) 'akhiye'p — (Sp) Romero

Various early sources state that wild *romero* was used by local Indians and Spanish Californians as a stimulant or tonic (Bingham 1890:37; Bard 1894:6). As an herb "for internal use by women," it may have been used in menstrual disorders (Birabent n.d.). The plant was regarded as a powerful disinfectant, valuable for livestock as well as humans, and an effective cure for gangrene (Birabent n.d.; Benefield 1951:21, 24, 26).

Harrington's consultants had heard that the wild *romero* plant, which grows in the mountains, was a powerful remedy. They did not know how it was used, however, and the Ventureño Chumash name they gave for the plant merely means "medicine." *Trichostema* was either not particularly important in Chumash medicine or had gone out of use before the turn of the century.

The common name "*romero*" is also applied to the cultivated garden rosemary, which may have supplanted the wild form as an herbal remedy. Contemporary Chumash and other Indian people in the Santa Barbara area are said to regard the wild herb as preferable to the domestic form. A tea of the leaves and flowers is drunk for stomachache, nervous troubles, and rheumatism; it is also used as a douche, as a rinse to darken and strengthen the hair, and as a wash for skin conditions or sores. The leaves are sometimes used as a flavoring agent in food (Weyrauch 1982:16).

WOOLLY BLUE CURLS
Trichostema lanatum

If a woman desired to induce an abortion during the first two months of pregnancy, she might drink a very hot tea of garden rosemary mixed with vinegar or wine and avoid exercising or going outdoors in cold weather (Juanita Centeno, personal communication 1978). The latter stipulation, a rare instance of "hot-cold" beliefs in the Chumash area, may indicate this use of rosemary was derived from Hispanic medicine. It is unknown whether the Chumash took the native plant *Trichostema* for this purpose.

The Luiseño gave a decoction of *romero* leaves to a woman who had "caught cold in the womb" and could not bear a child (Harrington 1933:194). The Ohlone used a related species, vinegar weed (*Trichostema lanceolatum*), to treat sores, colds, and stomachache (Bocek 1984:253).

Fabaceae | Pea Family

Trifolium spp. | **CLOVER**

(B) sha'puk' | (Sp) Tuche

(I) shapuk'

(V) shapuk

Chumash people ate the leaves of clover raw, like lettuce. After a winter of living on dried, stored food, fresh young clover shoots must have been a treat. Luisa Ygnacio of Santa Barbara remembered that when old María Ygnacio lay dying, she wanted to eat clover, and Luisa went out looking for it. In Santa Ynez, María Solares said that people ate the seeds "quite a bit." Many other California peoples ate clover either raw or cooked, and they particularly relished the fresh green leaves and shoots in springtime (Mead 2003:420).

Typhaceae | Cattail Family

Typha spp. | **CATTAIL**

(B, I) taqsh | (Sp) Tule Ancho

(V) khap

Three local species—narrow-leaved cattail (*Typha angustifolia* L.), southern cattail (*T. domingensis* Pers.), and broad-leaved cattail (*T. latifolia* L.)—were probably used interchangeably. The Spanish name, "wide tule," reflects the fact that cattail is similar to tule (*Scirpus* spp.) in habitat, appearance, and uses but has wider leaves. The Chumash clearly distinguished between these two unrelated plants.

A form of bread was made from cattail roots. The thick, starchy rhizomes were dug up, pulverized, and set out on a mat to dry, and balls of this dough were baked in hot ashes. This food was mentioned by some of Harrington's consultants who had seen or heard of it being made by the Yokuts. It may not have been common among the Chumash.

Chumash people boiled the whole cattail spikes like corn. They also ate cattail pollen before it was fully ripe by stirring it into water to make a thin, uncooked mush. Neither of these appears to have been a particularly important food source.

Cattail stems and leaves were used in the same way as tule stalks in thatching houses. According to Simplicio Pico, the Ventureño Chumash made twined mats of cattail leaves, but no specimens of this type are presently known.

The Ohlone were among the many other Indian peoples who ate cattail roots, young shoots, and pollen (Bocek 1984:255). In addition to these uses, cattail down was used for bedding in northern California (Chesnut 1902:310), and the stalks were made into mats by the Cahuilla (Bean and Saubel 1972:143).

Lauraceae — Laurel Family

Umbellularia californica (HOOK. & ARN.) NUTT.

CALIFORNIA BAY, CALIFORNIA LAUREL

(B, I) psha'n

(V) psha'an

(Sp) Laurél

The Chumash were noted for their fine woodworking. One example of a Chumash wooden bowl, now in the collections of the Santa Barbara Museum of Natural History, was made from the burl of a bay tree. The burl was hollowed out by burning and scraping, then polished and sealed with a mixture of red ochre and animal fat (Hudson 1977).

Before going out to hunt deer, a hunter burned leafy green laurel twigs and stood in the pungent smoke. The Chumash believed that deer liked this smell and were attracted to it. But it also made the animals dull and dizzy, hence easier to shoot. Presumably the strong scent of bay camouflaged the human smell of the hunter as well.

Bay leaves boiled in water were considered a strong remedy; this tea was drunk for colds. A headache remedy was made from bay leaves mixed with lard. Other medicinal uses included drinking bay leaf tea to treat diarrhea (Birabent n.d.) and putting the leaves in hot bath water to relieve rheumatism (Benefield 1951:24). Indian descendants nowadays tie bay leaves around their heads to cure headache, scatter them around their houses to repel fleas, and use them as a flavoring for food (Weyrauch 1982:6). Reportedly the leaves were also used to repel witches.

Sniffing crushed bay leaves can cause severe headache, and it is interesting that these leaves have been widely used as a headache cure. The other Chumash medicinal uses of the plant were also common elsewhere. In northern California, many Indian peoples ate bay fruits, which were commonly known as "peppernuts" (Mead 2003:430). The Ohlone ate the outer fruit pulp, either raw or boiled, and roasted the kernels or ground them into flour for cakes (Bocek 1984:249). There is no indication that the Chumash utilized this food source.

CHUMASH DEER HUNTERS

Imitating the movements of deer and disguised in stuffed decoy head-dresses, Chumash hunters could approach within close range of their quarry. To prepare themselves both physically and spiritually, the hunters first purified themselves in the sweat bath, washed in cold water, and then stood in the smoke of burning bay leaves. This treatment removed their human scent and was thought to disorient the deer, the better to ensure a successful hunt.

Urticaceae — Nettle Family

Urtica dioica L. subsp. *holosericea* (NUTT.) THORNE **GIANT CREEK NETTLE**

(B, I, V) khwapsh — (Sp) Ortiga

(Cr) qwap'sh

(O) tqmapsɨ

Although less important than Indian-hemp (*Apocynum*) or milkweed (*Asclepias*), nettle stems contain a fiber that the Chumash made into cordage. One reason for nettle's lesser value may have been that its fibers tend to end or weaken at leaf nodes, requiring more frequent splicing (Mathewson 1985:25–26). The Chumash only used long-jointed nettle stems for cordage. To test the fiber for strength, they held the ends behind their hands and gave three sharp jerks to see if it would break or pull apart. Fresh nettles are covered with stinging hairs but dried plants are not painful to handle.

Despite its disadvantages, nettle cordage was specified for a number of purposes. Nettle-string fishing lines consisted of one main line weighted with a stone, and branching off from this were eight or ten short lines, each with its own hook. Three-ply nettle string a quarter of an inch thick was twisted on the thigh to make the harpoon line. Considerable fiber was required, for the harpoon line was about two hundred and forty feet long. Nets made of nettle or Indian-hemp cordage were used in fishing for salmon or steelhead in rivers. The boards in plank canoes were held together with lashings of nettle or Indian-hemp cordage. A net for gathering acorns and wild cherry was made of nettle or milkweed in areas where Indian-hemp was scarce.

The bullroarer, a thin slat of wood whirled overhead to make a whirring sound, was attached to a string made of nettle. Dancing aprons were made of woodpecker or magpie feathers attached to nettle or milkweed threads. Fernando Librado said these skirts were worn by women and boys but not by men.

Nettle was also used medicinally. The Chumash treated paralysis and rheumatism by whipping the affected parts of the body with fresh nettle stems. According to María Solares, a rather dramatic alternative method could be employed to remedy the aches and pains of old age. She told Harrington: "The Indians here, both men and women, used to cut nettles and place them on the ground like a bed, lie down on them and writhe, thus treating the whole body until they began to sweat, and then they would go jump in the water. And they say it made them well, but yes, of course it hurt."

A similar technique was reported by a missionary at San Fernando in 1814: "When they suffer pains in the side they put red ants in water and apply them alive externally at the same time striking themselves with nettles" (Geiger and Meighan 1976:73). Indians in the Ventura area treated palsies by flagellation with nettles (Bard 1894:8). People at Santa Ynez remember older folks whipping themselves with nettle branches, and at least one person was still following this practice in the late 1970s (Weyrauch 1982:12).

The irritation induced by contact with stinging nettle increases the flow of blood to the affected area in much the same way as liniments and alcohol rubs, and it may actually promote rapid healing of damaged tissues. Among the other California peoples who utilized this treatment were the Cahuilla (Bean and Saubel 1972:143–144), Ohlone (Bocek 1984:250), and Pomo (Goodrich, Lawson, and Lawson 1980:77).

Urtica provided the only bast fiber of importance from non-latex-producing annual plants in the West, and it was used by nearly all Northwest Coast peoples (Hoover 1974:8–9). Nettles are common throughout California, but only the Modoc and Klamath held this fiber in high esteem. The Cahuilla, Gabrielino, Luiseño, Pomo, and Yokuts are among the other peoples who made nettle-fiber cordage (Mathewson 1985:25).

Verbenaceae — Verbena Family

Verbena lasiostachys LINK var. *lasiostachys* — **WESTERN VERVAIN**

(B) shikhwapsh 'i'ashk'a' — (Sp) Verbena

(I) s'uwmo' 'oyoso

(V) 'also'o, shikhwapsh 'i'ashk'á'

Verbena was also known locally by the Spanish terms *ortiga del coyote* ("coyote nettle") and *comida del jicote* ("bee food"). These are translations, respectively, of Barbareño and Ineseño Chumash descriptive names given by Harrington's consultants.

This plant was a well-known remedy for fever. The root was boiled and the decoction drunk, sometimes with sugar added to counter its bitter taste. Lucrecia García credited this tea with curing her of smallpox. Water in which verbena leaves had been crushed was strained and used in washing hair, in order to freshen the scalp.

The Spanish name "*verbena*" applies to *Verbena* spp. throughout northern Mexico and the Southwest. These plants are taken as a tea for stomach trouble, fever, and tuberculosis and used both in tea and as a wash for boils and wounds (Ford 1975:323). In California, the Ohlone used *V. lasiostachys* tea as a fever remedy (Bocek 1984:253), and the Concow ate *V. hastata* seeds as pinole (Chesnut 1902:383).

Vitaceae — Grape Family

Vitis californica BENTH. — **CALIFORNIA WILD GRAPE**

Vitis girdiana MUNSON — **DESERT WILD GRAPE**

(B) nu'nit' — (Sp) Uva Cimarrona

(I) nunit'

(V) nunit

Both these species have been found within Chumash territory, but it is not clear whether they are indigenous. In neighboring Kern County there are two Grapevine Canyons, one of which was named as early as 1806 (Gudde 1998:150). There is a Grapevine Creek in the western Tehachapi Mountains, on the edge of Chumash territory, where *V. californica* was once abundant; further east in the southern Sierra Nevada, *V. girdiana* grows in dense colonies in another Grapevine Canyon (Twisselmann 1967:281–282). A large grape thicket along the Grapevine Trail, west of Big Pine Mountain in Santa Barbara County, may have been planted in the nineteenth century, and grape colonies in Montecito and Ojai were probably also introduced in historic times (Smith 1976:190–191; 1998:370–371).

Harrington's Chumash consultants agreed that wild grapes were not found nearby but in the Tejon region. Chumash people picked and ate the sour fruit of the wild grape. They said the grapes did not belong to anyone, so it was all right to eat them. Fernando Librado spoke of a belief about wild grapes that may have been intended as a caution for humans as well:

> In the land of our Indian people there was nothing so strong as the bear. The Indians respected the world, and because the bear was strongest, they respected him too. The bear, although he is so strong, does have one thing which he must appreciate and respect—the wild grape, or nunit. The bear knows that if he eats too much of this fruit, it will ferment in his insides and cause him to become intoxicated. For this reason the bear has respect for the wild grape.
>
> —Fernando Librado (Hudson et al. 1977:81)

A Kitanemuk consultant said that old men used to pick wild grapes and tie them up in willow twigs. This kept the grapes from drying or spoiling and they lasted a long time that way. By the time Harrington interviewed this woman, though, people generally just let the wild grapes go to waste. The Ohlone ate fruit of *V. californica* raw (Bocek 1984:251); the Cahuilla ate desert grapes fresh, dried as raisins, and boiled (Bean and Saubel 1972:144). Grapevines were an important lashing and basketry material in northern California, where the plants are more common (Chesnut 1902:369; Goodrich, Lawson, and Lawson 1980:52).

Asteraceae — Sunflower Family

Xanthium strumarium L.

COCKLEBUR

(B) sho'moy
(I) mokoksh
(V) shomoy

(Sp) Guachapore, Cadillo

The Chumash names for this plant were applied to certain other spiny-fruited plants, particularly the introduced weed bur clover (*Medicago polymorpha*). The term *sho'moy* or *chomoy* (V) may also refer to edible greens in general.

Cocklebur leaves were boiled to make a wash for cuts. Santa Ynez Chumash people used this tea for bladder trouble (Gardner 1965:299). A priest at San Luis Obispo knew the same remedy (Garriga 1978:41). The decoction was also given to horses to drink for bladder inflammation, and the leaves were crushed and rubbed on the skin for ringworm (Birabent n.d.).

The Ohlone drank a tea of the plant for bladder ailments and ate the seeds in pinole (Bocek 1984:255). In Baja California and New Mexico, *Xanthium* was used medicinally to treat diarrhea and snakebite (Ford 1975:148).

Liliaceae — Lily Family

Yucca whipplei TORREY — **CHAPARRAL YUCCA, SPANISH BAYONET**

(B, I) pokh — (Sp) Maguey, Mescal, Quiote
(Cr, V) shtakuk
(O) ts'isuyi'
(additional terms given below)

Chumashan languages have extensive terminology for different parts of this plant in raw, cooked, fresh, and dry states. Local Spanish names for yucca are generally the same as those applied elsewhere to agave, a plant not found in Chumash territory until the mission padres introduced *Agave americana* from Mexico.

The following table shows the most common Chumashan and Spanish terms for yucca, as reconstructed from Harrington's notes.

Chumashan and Spanish Terms for Yucca

	Barbareño	Ineseño	Ventureño	Spanish
Plant	pokh	pokh	shtakuk	maguey (mescal, quiote)
Base of plant, esp. with leaves removed (also called a crown, head, or cabbage)				
Raw	pokh	pokh	wɨp	cabeza
Cooked	wɨp	wɨp	wɨp	mescal
Leaves	pokh	pokh	wɨp	?
Cooked	wɨp	wɨp	wɨp	mescal
Flower Stalk				quiote
Green	stakúk	stakuk	shtakuk	quiote
Cooked	wɨp	shnokhsh'en	shtakuk	mescal
Dry	meq	meq	chlumes	quiote
Growing tip or bud of inflorescence, esp. if cooked	?	shnokhsh'en	snokhsh'en	?

Among the Chumash, yucca was most important as a food plant, and it is frequently mentioned in the historical literature as well as in Harrington's notes. The summary log of the 1542 Cabrillo voyage by Paez noted that Chumash villagers at Rincon ate "maguey" and called maize "*oep*" (Wagner 1941:47–48). According to Harrington, "*oep*" was actually *wɨp*, a species of yucca known as *mezcal* (Sp.), the earth-roasted "cabbage" of which was "a sort of corn or staff of life to the people" (Harrington 1944:32).

In 1769 Pedro Fages, passing through the San Luis Obispo area, wrote:

> There is a great deal of century plant of the species which the Mexicans call mescali. The mode of using it is as follows: they make a hole in the ground, fill it in compactly with large firewood which they set on fire, and then throw on top a number of stones until the entire fire is covered but not smothered. When the stones are red hot, they place among them the bud of the plant; this they protect with grass or moistened hay, throwing on top a large quantity of earth, leaving it so for the space of twenty-four hours. The next day they take out their century plant roasted, or tlatemado [Sp. tatemado, "roasted"]. It is juicy, sweet, and of a certain vinous flavor; indeed a very good wine can be made from it (Fages 1937:50).

In January of 1770, Juan Crespí's party was treated to "roasted *mezcales*" when traveling through the Santa Monica Mountains (Crespí 1927:267). Since agave had not yet been introduced to this part of California by the Spanish, "*mescali*" in the quote refers to yucca, and the term "century plant" probably came from the translator. Fages's account of yucca preparation by the Obispeño Chumash is much like that provided by Harrington's Chumash consultants a century and a half later, which follows.

A pit two feet wide and three feet long was the right size to roast six or eight yucca crowns; some roasting pits were eight feet across and four feet deep. A fire was built in the pit and when it had burned down to coals, the leaves were burned off in the flames of another fire beside the pit. This singeing left just the thick succulent base of the plant, variously called the body, head, crown, bulb, or cabbage of the yucca. Rather than being burned away, the leaves could be cut off the plant and cooked too, but first they had to be trimmed and singed to remove spines and sharp margins.

The yucca crowns were placed four inches apart between layers of branches on top of the coals in the pit. If the leaves were to be cooked, they were put on before the crowns. The whole thing was covered with earth and left until the next day. It was important to let the roasted yucca cool completely, for if eaten warm, it induced diarrhea.

Fresh green yucca flower stalks could be cut into sections and prepared in the same manner or cooked in hot ashes, turning them periodically. When the skin was evenly burned, the stalks were done. They were set aside until cold and the skin was peeled or scraped off before eating. The interior was white and sometimes tasted slightly sweet or like bananas. Some stalks were more tender than others. Usually people ate the entire tip of the stalk, but they just chewed and sucked out the juice from the lower part, then spit out the fiber. A Chumash myth says that when Coyote's twenty-four children were born, he gathered up some thrown-away *maguey* cuds and moistened them again for the babies to suck on. Perhaps the Chumash did the same for their own infants.

Harrington's consultants did not think the Chumash ate yucca flowers or seeds. Fernando Librado said the yucca "root" was roasted and eaten like the stalk, but he may have meant the base or crown of the plant, not the actual root.

Yucca was not one of the principal Chumash fiber plants, but it was occasionally used to make cordage. The Chumash boiled the white central leaves and pounded them on rocks to obtain the fiber. Cordage makers would twist this fiber into string by rolling it on their thighs. Only the leaves were used to make cordage; fiber from the flower stalk was not suitable. Yucca fiber was relatively short in comparison to that obtained from century plants, which was used for cordage in historic times. Few of Harrington's consultants had seen yucca being made into string and some thought it was not used.

Fishing lines and possibly nets and men's belts were made from yucca string. "*Maguey* thread" was said to be used for sewing canoe planks. A headband found in a dry cave in the Cuyama area was woven of cordage identified as yucca fiber; the cordage is two-ply, S-twist strands two millimeters in diameter (Grant 1964:12).

Unspun yucca fiber was reportedly used to make sandals. Fernando Librado said the Chumash at Mugu soaked roots of yucca in water for about a month and then shredded them. They fashioned bundles of this material the length of the foot and sewed them together using a figure-eight stitch. People wore sandals to keep from blistering their feet when the ground was hot (Hudson and Blackburn 1985:97–98).

Yucca-fiber sandals, characterized as "imported," were listed as present among the Ventureño but not among the other Chumash groups (Harrington 1942:19).

Ear-piercing and tattooing were two forms of personal ornamentation in which yucca was used. As a girl, María Solares had her ears pierced with a yucca needle and fiber twisted on the thigh. The string was left tied on her earlobe for a week or more and her ears became infected. This method was used by the women of Tejon, she said. The Chumash more often pierced ears with a cactus thorn and put a stick in the hole. In tattooing, a cactus thorn was used to prick the skin, and charcoal from burned yucca heads was rubbed on for black designs.

The Chumash used mature, dry yucca flower stalks as tinder for starting fires. Historically, a spark struck with flint and steel was caught on the end of a section of yucca stalk and blown into flame. After each use the end was cut off, eventually using up the piece. The dry stalk may have been used in a similar way in earlier times, when fires were started with firesticks.

This species of yucca was utilized for food by several California peoples. Eugenia Mendez told Harrington the Kitanemuk burned off the yucca leaves before cutting the plant at ground level. If several men were roasting their harvest together, each put his yucca crowns in a single layer in the roasting pit and covered it with green herbage to keep it separate from the others. After cooking for two nights the crowns were red and sweet. The flower stalks were cooked out in the hills where they were gathered, not at home. In a method somewhat different from that of the Chumash, the Kitanemuk roasted the whole stalks in a blazing fire, turning them to cook evenly, then wiped off the layer of ashes. They cut the stalks into three-foot lengths and carried them home in sacks. They said the white interior of the cooked flower stalks had a sweet and distinctive taste different from the roasted crown.

The Cahuilla roasted the flower stalks of *Yucca whipplei* and ate the flowers fresh, boiled, or dried. They did not like this species for string-making because the fiber was stiff and sharp, and it cut the cordage maker's hands (Bean and Saubel 1972:150–151). The Luiseño ate *Yucca whipplei* heads, stalks, and flowers, but apparently they did not use the fiber (Sparkman 1908:234). The Diegueño ate stalks and flowers of this species and made cordage from the leaf fibers (Hedges and Beresford 1986:45).

Liliaceae | Lily Family

Zigadenus fremontii (TORR.) S. WATSON | **CHAPARRAL ZYGADENE**

Zigadenus venenosus S. WATSON | **DEATH CAMAS**

(B) mo'yoq' | (Sp) Cebolla Silvestre (see below)

(I) moyoq'

(V) moyoq

The Chumash collected zygadene in the mountains behind Santa Ynez and mashed the bulbs as a poultice for sores. It was said that sorcerers secretly gave a bit of this plant to people they wanted to kill; the victims died with blood coming from their mouths. Livestock became ill or died from eating zygadene leaves, which resemble those of soaproot (*Chlorogalum*).

These very poisonous members of the lily family have onion-like bulbs, and when Harrington's Chumash consultants talked about zygadene they often called it by the Spanish words for "wild onion." In most early literature, however, "wild onion" refers to the edible bulb of brodiaea (*Dichelostemma capitatum*). Obviously it is important not to confuse the two plants. Despite the toxicity of these bulbs, some California peoples reportedly did eat them after lengthy processing, and a number of groups used them medicinally (Strike 1994:169).

"Only an old-time mountain Indian
would know the names of all
the plants that grow in the mountains."

—Maria Solares

BIBLIOGRAPHY OF REFERENCES CITED

Ambris, Doroteo

1974 *Medicinal Herbs of Early Days in Use and Collected in the San Antonio Mission District, with Ambrisian-Latin Remedial and Common Index*, tr., ed. Valance E. Heinsen. Jolon, Calif.: Old Mission San Antonio de Padua.

Anderson, M. Kat

1996 "The Ethnobotany of Deergrass, *Muhlenbergia rigens* (Poaceae): Its Uses and Fire Management by California Indian Tribes." *Economic Botany* 50(4):409–422.

2005 *Tending the Wild: Native American Knowledge and the Management of California's Natural Resources*. Berkeley: University of California Press.

Applegate, Richard

1974 "Chumash Placenames." *Journal of California Anthropology* 1(2):187–205.

1975a "The Datura Cult among the Chumash." *Journal of California Anthropology* 2(1):7–17.

1975b "An Index of Chumash Placenames." In *Papers on the Chumash*, San Luis Obispo County Archaeological Society Occasional Paper no. 9:19–46. San Luis Obispo, Calif.

Armstrong, Wayne P.

1986 "The Deadly *Datura*." *Pacific Discovery* 39(4):35–41.

Baker, John R.

1994 "The Old Woman and Her Gifts: Pharmacological Bases of the Chumash Use of *Datura*." *Curare* 17(2):253–276.

Balls, Edward K.

1962 *Early Uses of California Plants*. California Natural History Guides 10. Los Angeles: University of California Press.

Barber, Edward

1931 Unpublished correspondence to Ralph Hoffmann, transcribed copies on file at Santa Barbara Museum of Natural History.

Bard, Cephas L.

1894 "A Contribution to the History of Medicine in Southern California." *Southern California Practitioner*, August 1894:3–29.

1930 "Medicine and Surgery among the First Californians." *Touring Topics*, January 1930:20–30.

Barrett, S. A., and E. W. Gifford

1933 "Miwok Material Culture: Indian Life of the Yosemite Region." *Bulletin of the Milwaukee Public Museum* 2(4):116–377. Yosemite National Park, Calif.: Yosemite Natural History Association.

Barrows, David Prescott

1900 *The Ethno-Botany of the Coahuilla Indians.* Chicago: University of Chicago Press. Reprint 1978, New York: AMS Press.

Bean, Lowell John and Katherine Siva Saubel

1972 *Temalpakh: Cahuilla Indian Knowledge and Uses of Plants.* Banning, Calif.: Malki Museum Press.

Beauchamp, Philip S., Albert T. Bottini, Vasu Dev, Anand B. Melkani, and Jan Timbrook

1993 "Analysis of the Essential Oil from *Lomatium californicum* (Nutt.) Math. and Const." In *Food Flavors, Ingredients and Composition,* ed. G. Charambolous. New York: Elsevier Science BV: pp. 605–610.

Benefield, Hattie S.

1951 *For the Good of the Country: The Life Story of William Benjamin Foxen.* Los Angeles: Lorrin L. Morrison.

Bingham, Mrs. R. F.

1890 "Medicinal Plants Growing Wild in Santa Barbara and Vicinity." *Santa Barbara Society of Natural History Bulletin* 1(3):34–37.

Birabent, Frank L.

n.d. "Wild Herbs Used by the Indians and Spanish People in Santa Barbara County for Medicinal Purposes." Unpublished manuscript on file, Santa Barbara Museum of Natural History.

Blackburn, Thomas C.

1963 "A Manuscript Account of the Ventureño Chumash." *Archaeological Survey Annual Report 1962–1963.* Los Angeles: University of California, pp. 135–158.

1975 *December's Child: A Book of Chumash Oral Narratives.* Berkeley: University of California Press.

Blochman, Ida May

1893a "Californian Herb-Lore I." *Erythea* 1(9):190–191.

1893b "Californian Herb-Lore II." *Erythea* 1(11):231–233.

1894a "Californian Herb-Lore III." *Erythea* 2(1):9–10.

1894b "Californian Herb-Lore IV." *Erythea* 2(3):39–40.

1894c "Californian Herb-Lore V." *Erythea* 2(10):162–163.

Bocek, Barbara R.

1984 "Ethnobotany of Costanoan Indians, California, Based on Collections by John P. Harrington." *Economic Botany* 38(2):240–255.

Bolton, Herbert E. (ed.)

1925 *Spanish Exploration in the Southwest 1542–1706.* New York: Charles Scribner's Sons.

Boscana, Fr. Gerónimo

1933 *Chinigchinich: A Revised and Annotated Version of Alfred Robinson's Translation of Father Gerónimo Boscana's Historical Account of the Belief, Usages, Customs and Extravagancies of the Indians of this Mission of San Juan Capistrano Called the Acagchemem Tribe*, tr. Alfred Robinson, annotated by John P. Harrington. Santa Ana, Calif.: Fine Arts Press. Reprint 1978, Banning, Calif.: Malki Museum Press.

Braje, Todd J., Jon M. Erlandson, and Jan Timbrook

2005 "An Asphaltum Coiled Basket Impression, Tarring Pebbles, and Middle Holocene Water Bottles from San Miguel Island, California." *Journal of California and Great Basin Anthropology* 25(2):207–213.

Brand, Rochelle, and Walden Townsend

1978 "Herb and Plant Remedies Used in the Santa Barbara Area." Unpublished manuscript sponsored by the Santa Barbara Urban Indian Health Project, on file at Santa Barbara Museum of Natural History.

Brown, Alan K.

1967 *The Aboriginal Population of the Santa Barbara Channel.* Reports of the University of California Archaeological Survey, no. 69. Berkeley: University of California Archaeological Research Facility, Department of Anthropology.

Callaghan, Catherine A.

1975 "J. P. Harrington—California's Great Linguist." *Journal of California Anthropology* 2(2):183–187.

Chesnut, V. K.

1902 "Plants Used by the Indians of Mendocino County, California." *Contributions of the U.S. National Herbarium* 7:285–422. Reprint 1974, Fort Bragg, Calif.: Mendocino County Historical Society.

Cook, Sherburne F.

1960 *Colonial Expeditions to the Interior of California: Central Valley, 1800–1820.* Anthropological Records 16(6). Berkeley: University of California Press.

Coville, Frederick V.

1904 "Plants Used in Basketry." In *Aboriginal American Basketry*, by Otis T. Mason, Annual Report of the Smithsonian Institution for 1902. Washington, D.C.: U.S. Government Printing Office, pp.199–214.

Craig, Steve

1966 "Ethnographic Notes on the Construction of Ventureño Chumash Baskets from the Ethnographic and Linguistic Field Notes of John P. Harrington." *Archaeological Survey Annual Report* 8:197–214. Los Angeles: University of California.

1967 "The Basketry of the Ventureño Chumash." *Archaeological Survey Annual Report* 9:78–149. Los Angeles: University of California.

Crespí, Juan

1927 *Fray Juan Crespí, Missionary Explorer on the Pacific Coast, 1769–1774*, ed. Herbert E. Bolton. Berkeley: University of California Press.

Cunningham, Richard W.

1989 *California Indian Watercraft.* San Luis Obispo, Calif.: EZ Nature Books.

Curtin, L. S. M.

1965 *Healing Herbs of the Upper Rio Grande.* Los Angeles: Southwest Museum.

Dawson, Lawrence, and James Deetz

1965 "A Corpus of Chumash Basketry." *Archaeological Survey Annual Report* 7:193–275. Los Angeles: University of California.

Earle, F. R., and Quentin Jones

1962 "Analyses of Seed Samples from 113 Plant Families." *Economic Botany* 16(4):221–250.

Elmore, Francis H.

1943 *Ethnobotany of the Navajo.* Monograph series 1(7). Albuquerque, N.M.: University of New Mexico Press.

Fages, Pedro

1937 *A Historical, Political, and Natural Description of California by Pedro Fages, Soldier of Spain, Dutifully Made for the Viceroy in the Year 1775*, tr. Herbert J. Priestley. Berkeley: University of California Press.

Farris, Glenn

1980 "A Reassessment of the Nutritional Value of *Pinus monophylla.*" *Journal of California and Great Basin Anthropology* 2(1):132–136.

1982 "Pine Nuts as an Aboriginal Food Source in California and Nevada: Some Contrasts." *Journal of Ethnobiology* 2(2):114–122.

Felger, Richard S., and Mary Beck Moser

1976 "Seri Indian Food Plants: Desert Subsistence without Agriculture." *Ecology of Food and Nutrition* 5(1):13–27.

Flora of North America Committee (eds.)

1993+ *Flora of North America North of Mexico.* 12+ vols. New York and Oxford: Oxford University Press.

Ford, Karen Cowan

1975 *Las Yerbas de la Gente: A Study of Hispano-American Medicinal Plants.* University of Michigan Museum of Anthropology Anthropological Papers no. 60. Ann Arbor, Mich.

Fuller, Thomas C., and Elizabeth McClintock

1986 *Poisonous Plants of California.* California Natural History Guides 53. Berkeley: University of California Press.

Gardner, Louise

1965 "The Surviving Chumash." *Archaeological Survey Annual Report* 7:277–302. Los Angeles: University of California.

Garriga, Andreas

1978 *Andrew Garriga's Compilation of Herbs & Remedies Used by the Indians & Spanish Californians Together with Some Remedies of His Own Experience*, ed. Msgr. Francis J. Weber. Los Angeles: Archdiocese of Los Angeles.

Geiger, Maynard, and Clement W. Meighan

1976 *As the Padres Saw Them: California Indian Life and Customs as Reported by the Franciscan Missionaries, 1813–1815*. Santa Barbara Bicentennial Historical Series no. 1. Santa Barbara, Calif.: Santa Barbara Mission Archive Library.

Gifford, Edward W.

1967 *Ethnographic Notes on the Southwestern Pomo*. Anthropological Records 25:1–47. Berkeley: University of California Press.

Gilliland, Linda E.

1985 "Proximate Analysis and Mineral Composition of Traditional California Native American Foods." Unpublished master's thesis, University of California, Davis.

Goodrich, Jennie, Claudia Lawson, and Vana Parrish Lawson

1980 *Kashaya Pomo Plants*. American Indian Monograph Series no. 2. Los Angeles: American Indian Studies Center, University of California.

Grae, Ida

1974 *Nature's Colors: Dyes from Plants*. New York: Macmillan.

Grant, Campbell

1964 "Chumash Artifacts Collected in Santa Barbara County, California." *Reports of the University of California Archaeological Survey* 63:1–44. Berkeley: University of California Archaeological Research Facility.

1965 *The Rock Paintings of the Chumash*. Berkeley: University of California Press.

Greenwood, Roberta S.

1972 "9000 Years of Prehistory at Diablo Canyon, San Luis Obispo County, California." San Luis Obispo County Archaeological Society Occasional Paper no. 7. San Luis Obispo, Calif.

Gudde, Erwin G.

1998 *California Place Names: The Origin and Etymology of Current Geographical Names*, 4th ed., rev., enlarged by William Bright. Berkeley: University of California Press.

Harrington, John P.

1928 "The Mission Indians of California." In *Explorations and Field-Work of the Smithsonian Institution in 1927*, Smithsonian Miscellaneous Collections, pub. 2957:173–178. Washington, D.C.: Smithsonian Institution.

1933 Annotations. In *Chinigchinich: A Revised and Annotated Version of Alfred Robinson's Translation of Father Gerónimo Boscana's Historical Account of the Belief, Usages, Customs and Extravagancies of the Indians of this Mission of San Juan Capistrano Called the Acagchemem Tribe*, tr. Alfred Robinson. Santa Ana, Calif.: Fine Arts Press. Reprint 1978, Banning, Calif.: Malki Museum Press.

1935 "Field-work among the Indians of California." *Explorations and Field-Work of the Smithsonian Institution in 1934*, pp. 81–84. Washington, D.C.: Smithsonian Institution.

1942 *Culture Element Distributions XIX: Central California Coast.* Anthropological Records 7(1). Berkeley: University of California Press.

1944 "Indian Words in Southwest Spanish, Exclusive of Proper Nouns." *Plateau* 17(2):27–40.

1986 *Native American History, Language and Culture of Southern California/Basin.* Papers of John P. Harrington in the Smithsonian Institution 1907–1957, vol. 3, ed. Elaine L. Mills and Ann J. Brickfield. Washington, D.C.: Smithsonian Institution, National Anthropological Archives. (Microfilm edition, Millwood, N.Y.: Kraus International Publications.)

1994 *The Photograph Collection of John Peabody Harrington in the National Anthropological Archives, Smithsonian Institution*, ed. Gerianne Schaad. Washington, D.C.: Smithsonian Institution, National Anthropological Archives.

Hedges, Ken, and Christina Beresford

1986 *Santa Ysabel Ethnobotany.* San Diego Museum of Man Ethnic Technology Notes no. 20. San Diego, Calif.

Heizer, Robert F.

1970 "More J. P. Harrington Notes on Ventureño Chumash Basketry and Culture." *Contributions of the University of California Archaeological Research Facility* no. 9:59–73. Berkeley: University of California, Department of Anthropology.

1973 *Notes on the McCloud River Wintu and Selected Excerpts from Alexander S. Taylor's Indianology of California*, Part II, pp. 24–79. Berkeley: University of California Archaeological Research Facility, Department of Anthropology.

Heizer, Robert F., and Albert B. Elsasser

1980 *The Natural World of the California Indians.* California Natural History Guides 46. Berkeley: University of California Press.

Henshaw, Henry W.

1885 "The Aboriginal Relics called 'Sinkers' or 'Plummets.'" *American Journal of Archaeology* 1(2):105-114.

1887 "Perforated Stones from California." *Bureau of American Ethnology Bulletin* 2:1–35. Washington, D.C.: U.S. Government Printing Office.

1955 *California Indian Linguistic Records: The Mission Indian Vocabularies of H. W. Henshaw*," ed. Robert F. Heizer. Anthropological Records 15(2). Berkeley: University of California Press.

Hickman, James C. (ed.)

1993 *The Jepson Manual: Higher Plants of California.* Berkeley: University of California Press.

Hitchcock, A. S.

1950 *Manual of the Grasses of the United States,* 2d rev. ed. U.S. Department of Agriculture Miscellaneous Publication no. 200. Washington, D.C.: U.S. Government Printing Office.

Hoover, Robert L.

1971 "Some Aspects of Santa Barbara Channel Prehistory." Ph.D. dissertation, University of California, Berkeley.

1973a *Chumash Fishing Equipment.* San Diego Museum of Man Ethnic Technology Notes no. 9. San Diego, Calif.

1973b "Salt and the California Indians." *Pacific Discovery* 26(2):25–28.

1974 "Aboriginal Cordage in Western North America." Imperial Valley College Museum Occasional Paper no. 1. El Centro, Calif.

Hudson, Travis

1974 *Chumash Archery Equipment.* San Diego Museum of Man Ethnic Technology Notes no. 13. San Diego, Calif.

1976 "Chumash Hinged-Stick Snares." *Masterkey* 50(4):124–137.

1977 *Chumash Wooden Bowls, Trays and Boxes.* San Diego Museum Papers no. 13. San Diego, Calif.: San Diego Museum of Man.

1979 *Breath of the Sun: Life in Early California as Told by a Chumash Indian, Fernando Librado, to John P. Harrington.* Banning, Calif: Malki Museum Press.

Hudson, Travis, and Thomas C. Blackburn

1982 *The Material Culture of the Chumash Interaction Sphere, vol. I, Food Procurement and Transportation.* Ballena Press Anthropological Papers no. 25. Los Altos, Calif.: Ballena Press; Santa Barbara, Calif.: Santa Barbara Museum of Natural History.

1983 *The Material Culture of the Chumash Interaction Sphere, vol. II, Food Processing and Shelter.* Ballena Press Anthropological Papers no. 27. Menlo Park, Calif.: Ballena Press; Santa Barbara, Calif.: Santa Barbara Museum of Natural History.

1985 *The Material Culture of the Chumash Interaction Sphere, vol. III, Clothing, Ornamentation, and Grooming.* Ballena Press Anthropological Papers no. 28. Menlo Park, Calif.: Ballena Press; Santa Barbara, Calif.: Santa Barbara Museum of Natural History.

1986 *The Material Culture of the Chumash Interaction Sphere, vol. IV, Ceremonial Paraphernalia, Games, and Amusements.* Ballena Press Anthropological Papers no. 30. Menlo Park, Calif.: Ballena Press; Santa Barbara, Calif.: Santa Barbara Museum of Natural History.

1987 *The Material Culture of the Chumash Interaction Sphere, vol. V, Manufacturing Processes, Metrology, and Trade.* Ballena Press Anthropological Papers no. 31. Menlo Park, Calif.: Ballena Press; Santa Barbara, Calif.: Santa Barbara Museum of Natural History.

Hudson, Travis, Thomas C. Blackburn, Rosario Curletti, and Jan Timbrook

1977 *The Eye of the Flute: Chumash Traditional History and Ritual as Told by Fernando Librado Kitsepawit to John P. Harrington.* Santa Barbara Bicentennial Historical Series no. 4. Santa Barbara, Calif.: Santa Barbara Museum of Natural History.

Hudson, Travis, and Jan Timbrook

1997 *Chumash Indian Games.* Santa Barbara, Calif.: Santa Barbara Museum of Natural History.

Hudson, Travis, Jan Timbrook, and Melissa Rempe

1978 *Tomol: Chumash Watercraft as Described in the Ethnographic Notes of John P. Harrington.* Ballena Press Anthropological Papers no. 9. Socorro, N.M.: Ballena Press.

Jepson, Willis Linn

1909 *A Flora of California*, Part II: Salicaceae to Urticaceae. San Francisco: Cunningham, Curtiss & Welch.

1914 *A Flora of California*, Part IV: Platanaceae to Portulacaceae. San Francisco: H. S. Crocker Co.

1925 *A Manual of the Flowering Plants of California.* Berkeley: University of California Press.

1936 *A Flora of California*, Vol. II: Capparidaceae to Cornaceae. San Francisco: California School Book Depository.

1939 *A Flora of California*, Vol. III, Part 1. Berkeley, Calif.: Associated Students Store.

1943 *A Flora of California*, Vol. III, Part 2. Berkeley, Calif.: Associated Students Store.

Junak, Steve, Tina Ayers, Randy Scott, Dieter Wilken, and David Young

1995 *A Flora of Santa Cruz Island.* Santa Barbara, Calif.: Santa Barbara Botanic Garden.

Kay, Margarita Artschwager

1996 *Healing with Plants in the American and Mexican West.* Tucson, Ariz.: University of Arizona Press.

King, Linda

1969 "The Medea Creek Cemetery (LAn-243): An Investigation of Social Organization from Mortuary Practices." *Archaeological Survey Annual Report* 11:23–68. Los Angeles: University of California.

Krochmal, Arnold, and Connie Krochmal

1974 *The Complete Illustrated Book of Dyes from Natural Sources.* Garden City, N.Y.: Doubleday.

Kroeber, Alfred L.

1908 *A Mission Record of the California Indians: From a Manuscript in the Bancroft Library.* University of California Publications in American Archaeology and Ethnology 8(1):1–27. Berkeley: University of California Press.

Laird, Carobeth

1975 *Encounter with an Angry God: Recollections of My Life with John Peabody Harrington.* Banning, Calif.: Malki Musem Press.

Latta, Frank F.

1977 *Handbook of Yokuts Indians,* 2d. ed. Santa Cruz, Calif.: Bear State Books.

Lawton, Harry W., Philip J. Wilke, Mary DeDecker, and William J. Mason

1976 "Agriculture among the Paiute of Owens Valley." *Journal of California Anthropology* 3(1):13–50.

Longinos Martínez, José

1961 *The Journal of José Longinos Martínez: Notes and Observations of the Naturalist of the Botanical Expedition in Old and New California and the South Coast, 1791–1792,* tr., ed. Lesley B. Simpson. San Francisco: Santa Barbara Historical Society/John Howell.

Martínez, Maximino

1979 *Catálogo de Nombres Vulgares y Científicos de Plantas Mexicanas.* Mexico City: Fondo de Cultura Económica.

Mathewson, Margaret S.

1985 "Threads of Life: Cordage and Other Fibers of the California Tribes." Unpublished senior thesis, Board of Studies in Anthropology, University of California, Santa Cruz.

Mead, George R.

1972 *The Ethnobotany of the California Indians: A Compendium of the Plants, Their Users, and Their Uses.* Occasional Publications in Anthropology, Ethnology Series no. 30. Greeley, Colo.: Museum of Anthropology, University of Northern Colorado.

2003 *The Ethnobotany of the California Indians.* La Grande, Ore.: E-Cat Worlds.

Merrill, Ruth Earl

1923 *Plants Used in Basketry by the California Indians.* University of California Publications in American Archaeology and Ethnology 20:215–242. Reprint 1973, Ramona, Calif.: Ballena Press.

Miksicek, Charles H.

1998 "Yellow Nut-grass (*Cyperus esculentus* L.): An Under-recognized Traditional California Food Plant." In *Plant Processing on the San Antonio Terrace: Archaeological Investigations at CA-SBA-2767,* by C. G. Lebow and D. R. Harro. Prepared by Applied Earthworks, Inc., for the Central Coast Water Authority, Buellton, Calif., pp. A.1–A.10.

Muenscher, Walter Conrad

1975 *Poisonous Plants of the United States.* New York: Collier Books.

Munz, Philip A., in collaboration with David D. Keck

1959 *A California Flora.* Berkeley: University of California Press.

Niethammer, Carolyn

1974 *American Indian Food and Lore.* New York: Collier Books.

O'Neale, Lila M.

1932 *Yurok-Karok Basket Weavers.* University of California Publications in American Archaeology and Ethnology 32(1):1–184.

Orr, Phil C.

1968 *Prehistory of Santa Rosa Island.* Santa Barbara, Calif.: Santa Barbara Museum of Natural History.

Padilla, Victoria

1961 *Southern California Gardens: An Illustrated History.* Berkeley: University of California Press.

Palmer, Edward

1878 "Plants Used by the Indians of the United States." *American Naturalist* 12:539–606, 646–655.

Rogers, David Banks

1929 *Prehistoric Man of the Santa Barbara Coast.* Santa Barbara, Calif.: Santa Barbara Museum of Natural History.

Romero, John Bruno

1954 *The Botanical Lore of the California Indians.* New York: Vantage Press.

Rothrock, J. T.

1876 "Report upon the Operations of a Special Natural-History Party and Main Field Party No. 1, California Section, Field Season of 1875, Being the Results of Observations upon the Economic Botany and Agriculture of Portions of Southern California." Appendix H-5, pp. 202–213 in *Annual Report upon Geographical Surveys West of the One Hundredth Meridian*...by George M. Wheeler, Appendix JJ, *Annual Report of the Engineers*, 1876. Washington, D.C.: Government Printing Office.

1878 "Notes on Economic Botany." In *United States Geographical Surveys West of the One Hundredth Meridian, in charge of George M. Wheeler*, vol. 6: Botany, pp. 39–52. Washington, D.C.: U.S. Government Printing Office.

Saunders, Charles Francis

1914 *With the Flowers and Trees in California.* New York: Robert McBride & Co. (third printing, 1923).

1920 *Useful Wild Plants of the United States and Canada.* New York: Robert M. McBride & Co.

1933 *Western Wildflowers and Their Stories.* Garden City, N.Y.: Doubleday, Doran, & Co.

Smith, Clifton F.

1976 *A Flora of the Santa Barbara Region, California.* Santa Barbara, Calif.: Santa Barbara Museum of Natural History.

1998 *A Flora of the Santa Barbara Region, California,* 2d ed. Santa Barbara, Calif.: Santa Barbara Botanic Garden/Capra Press.

Sparkman, Philip S.

1908 *The Culture of the Luiseño Indians.* University of California Publications in American Archaeology and Ethnology 8(4):187–234. Berkeley: University of California Press.

Stirling, Matthew

1963 "John Peabody Harrington." *American Anthropologist* 65(2):370–381.

Strike, Sandra S.

1994 *Ethnobotany of the California Indians,* vol. 2: *Aboriginal Uses of California's Indigenous Plants.* Champaign, Ill.: Koeltz Scientific Books USA.

Timbrook, Jan

1980 "A Wooden Artifact from Santa Cruz Island." *Journal of California and Great Basin Anthropology* 2(2):272–279.

1982 "Use of Wild Cherry Pits as Food by the California Indians." *Journal of Ethnobiology* 2(2):162–176.

1984 "Chumash Ethnobotany: A Preliminary Report." *Journal of Ethnobiology* 4(2):141–169.

1986 "Chia and the Chumash: A Reconsideration of Sage Seeds in Southern California." *Journal of California and Great Basin Anthropology* 8(1):50–64.

1987 "Virtuous Herbs: Plants in Chumash Medicine." *Journal of Ethnobiology* 7(2):171–180.

1990 "Ethnobotany of Chumash Indians, California, Based on Collections by John P. Harrington." *Economic Botany* 44(2):236–253.

1993 "Island Chumash Ethnobotany." In *Archaeology on the Northern Channel Islands of California: Studies of Subsistence, Economics, and Social Organization,* ed. Michael A. Glassow. Archives of California Prehistory no. 34:47–62. Salinas, Calif.: Coyote Press.

2000 "Search for the Source of the Sorcerers' Stones." *Proceedings of the Fifth California Islands Symposium,* OCS Study, MMS-99-0038. U.S. Department of the Interior, Minerals Management Service, Pacific OCS Region.

Timbrook, Jan, John R. Johnson, and David D. Earle

1982 "Vegetation Burning by the Chumash." *Journal of California and Great Basin Anthropology* 4(2):163–186.

Tulloch, Alice

1990 "Ethnobotany of Willow Basketry." Paper presented at Sixth California Indian Conference, University of California, Riverside, October 26, 1990.

Twisselmann, Ernest C.
1967 *A Flora of Kern County, California*. San Francisco, Calif.: University of San Francisco (reprinted from *The Wasmann Journal of Biology* 25(1) and 25(2), 1967).

Vizcaíno, Juan
1959 *The Sea Diary of Fr. Juan Vizcaíno to Alta California, 1769*, tr. Arthur B. Woodward. Early California Travels Series vol. 49. Los Angeles: Glen Dawson.

Voegelin, Erminie W.
1938 "Tubatulabal Ethnography." *Anthropological Records* 2(1):1–90. Berkeley: University of California Press.

Wagner, Henry R.
1929 *Spanish Voyages to the Northwest Coast of America in the Sixteenth Century*. California Historical Society Special Publication no. 4. San Francisco, Calif.
1941 *Juan Rodríguez Cabrillo: Discoverer of the Coast of California*. San Francisco: California Historical Society.

Walker, Phillip L., and Travis Hudson
1993 *Chumash Healing: Changing Health and Medical Practices in an American Indian Community*. Banning, Calif.: Malki Museum Press.

Webb, Edith Buckland
1952 *Indian Life at the Old Missions*. Lincoln, Neb.: University of Nebraska Press.

Weyrauch, Rita
1982 "Herbal Remedies." *Solstice Journal* 1(1):1–27. Santa Barbara Indian Center.

Wolf, Carl B.
1945 *California Wild Tree Crops: Their Crop Production and Possible Utilization*. Orange County, Calif.: Rancho Santa Ana Botanic Garden.

Woodward, Lisa L., and Martha J. Macri
2005 "J. P. Harrington Database Project: An Archival Resource for Anthropologists, Archaeologists, and Native Communities." *Journal of California and Great Basin Anthropology* 25(2):235–239.

Yarrow, H. C.
1879 "Report on the Operations of a Special Party for Making Ethnological Researches in the Vicinity of Santa Barbara..." In "Reports upon Archaeological and Ethnological Collections from the Vicinity of Santa Barbara, California," ed. Frederick W. Putnam. *United States Geographical Surveys West of the One Hundredth Meridian, in charge of George M. Wheeler* vol. 7, part 1, pp. 32–46. Washington, D.C.: U.S. Government Printing Office.

Yates, Lorenzo G.

n.d. "Interview with Rafael Solares and Justo Gonzales on 'charmstones.'" Undated handwritten manuscript, Lorenzo G. Yates Papers, Bancroft Library Box 2, C-B 472.85. Transcription by Travis Hudson, 1981, on file at Santa Barbara Museum of Natural History.

1890 "Charmstones: Notes on the So-called 'Plummets' or 'Sinkers.'" *Bulletin of the Santa Barbara Society of Natural History* 1(2):13–28.

1891 "Fragments of the History of a Lost Tribe. *American Anthropologist* 4(4):373–376.

Zigmond, Maurice L.

1941 "Ethnobotanical Studies among California and Great Basin Shoshoneans." Ph.D. dissertation. New Haven, Conn: Yale University.

1981 *Kawaiisu Ethnobotany.* Salt Lake City: University of Utah Press.

CHUMASHAN LANGUAGE TERMS

John P. Harrington, upon whose field notes this work is largely based, was first and foremost a linguist. In repeated interviews with each of his Chumash consultants, he took pains to transcribe words exactly as they were pronounced. Naturally there were differences in pronunciation between individual speakers, and sometimes even in the same individual from one day to the next. Also, during his long career, Harrington changed some of the symbols he used to represent particular sounds. These factors account for the many variations and inconsistencies in rendering what had been exclusively spoken languages.

Notes on Pronunciation

Vowels

- a as in 'father'
- e as in 'they'
- i as in 'machine'
- o as in 'note'
- u as in 'rude'
- ɨ (barred i) is a distinct vowel midway between short 'i' in 'tick' and short 'u' as in 'tuck', sometimes written elsewhere as 'ə' (schwa)

Consonants are like those in English, with the following exceptions:

- ' (glottal stop) is a catch in the throat, as in English 'uh-oh'
- č is pronounced 'ch' as in 'church'
- h (superscript 'h') is a more faintly aspirated sound
- q is like 'k' but farther back in the throat
- ɬ (barred l) is a sort of whispered, shushing 'l'
- š is pronounced 'sh' as in 'ship'
- x is breathy and harsh, like 'ch' in the German pronunciation of 'Bach,' here rendered as 'kh' in popularized spelling used in this book

Notes on Alphabetizing

- ' (glottal stop) counts as a letter, comes before 'a' in the alphabet
- ɨ counts as a letter, comes after 'i' in the alphabet

Popularized Spelling	Linguistic Spelling	Scientific Name	Common Name
'akhiye'p (V)	'axiye'p, 'axiya'ep ('medicine')	*Rosmarinus officinalis*	ROSEMARY
		Trichostema lanatum	WOOLLY BLUE CURLS
'akhtakuy (B, I)	'aqtakuy	conical burden basket	basket type
'akhtatapɨsh (V)	'axtatapɨš	*Prunus ilicifolia*	HOLLY-LEAVED CHERRY
'akhtayukhash (B, I)	'axtayuxaš	*Prunus ilicifolia*	HOLLY-LEAVED CHERRY
'akhwayɨsh (I)	'axwayɨš	*Chenopodium californicum*	SOAP PLANT
'akmila'ash (V)	'akmila'aš	drinking cup	basket type
'akupusha'sh (I)	'aKupušaš, 'aKupuša'š	drinking cup	basket type
'aku'w (P)	'aqu'w	*Quercus agrifolia*	COAST LIVE OAK
'alakhpɨy (V)	'alaxpɨy (descriptive: 'sour')	*Rumex hymenosepalus*	CANAIGRE
'alamakhwak'ay (I)	'alamaxwak'ay	*Clematis* spp.	CLEMATIS
'alakhnipk'ɨsh (V)	'alaxnipk'ɨš, 'alaxnipkɨy (descriptive: 'the flavor is sour')	*Rumex crispus*	CURLY DOCK
'alaqtaha (V)	'alaqtaha	*Mentha* spp.	MINT
		Satureja douglasii	YERBA BUENA
'alashkhalalash (B)	'alašxalalaš, 'alašxalalač	*Acourtia microcephala*	SACAPELLOTE
'alats'ɨmɨmɨ (I)	'alats'ɨmɨmɨ	narrow-mouthed basket	basket type
'alikhshanuch (V)	'alixšanuč (descriptive: 'ashy')	*Salvia apiana*	WHITE SAGE
'almakhmal 'i suninakhshep (V)	'almaxmal 'isuninaxšep	*Calystegia macrostegia*	MORNING-GLORY
'alsho'o (V)	'ałšo'o	*Verbena lasiostachys*	WESTERN VERVAIN
'alshuklash	'ałšuklaš	pipe doctor, a type of shaman	ritual practitioner
'aluche'esh (B)	'aluče'eš	*Avena* spp.	WILD OAT
'aluqchahay 'isakhpilil (V)	'aluqčahay 'i saxpilil	*Datisca glomerata*	DURANGO ROOT
'an (V)	'an	*Eriogonum elongatum, E. nudum*	WILD BUCKWHEAT
'anmakhwaka'y (V)	'anmaxwaka'y, 'anawaka'y, 'alamaxwák'ay, 'anmaxwaks'ay	*Marah macrocarpus*	WILD CUCUMBER
'ansiwa'wu'y (B)	'ansiwa'wú'y	*Datisca glomerata*	DURANGO ROOT
'aqnipshkáy (B)	'aqnipškáy	*Oxalis albicans*	WOOD-SORREL
'aqsho' (P)	'aqšo'	*Platanus racemosa*	WESTERN SYCAMORE

'aqulpop' (B, I, P)	'aquɨpop', 'aquɨpóp'	*Solanum douglasii*	DOUGLAS NIGHTSHADE
'awa'q (B, I)	'awa'q, 'awaq	water bottle	basket type
'ayakuy (I)	'ayakuy	winnowing tray	basket type
'ayip	'ayip	magical poison	ritual item
'ayuhat (V)	'ayuhat	winnowing tray	basket type
'ekhpe'w (B)	'expe'w	*Phragmites australis*	CARRIZO GRASS
'ekhpew (I)	'expew	*Phragmites australis*	CARRIZO GRASS
'epsu (B, I, V)	'epsu, 'ep'su	basket cap	basket type
'esmu (V)	'esmu	*Juncus acutus*	SPINY RUSH
		Juncus effusus	BOG RUSH
'i'laq (B)	'i'laq	*Croton californicus*	CALIFORNIA CROTON
'i'laq' (B)	'i'laq'	*Atriplex* spp.	SALTBUSH
'i'lepesh (I)	'i'lepeš	*Salvia columbariae*	CHIA
'ikhpanɨsh (B, I, V)	'ixpaniš, 'ixpánis	*Quercus* spp.	acorns
'ilepesh (B)	'ilépeš, 'i'lepeč	*Salvia columbariae*	CHIA
'imolɨ'ɨsh (V)	'imolɨ'ɨš	parching tray	basket type
'iqma'y (V)	'iqma'y	*Lepidium nitidum*	PEPPERGRASS
'itepesh (V)	'itepeš, 'it'epiš	*Salvia columbariae*	CHIA
'okhoy (V)	'oxoy, 'oxo'y, 'oqoi	gathering basket	basket type
'okhponush (B, I)	'oxpónuš, 'oxponuš	*Asclepias* spp.	MILKWEED
'onchochi (B, I)	'ončoči	*Anemopsis californica*	YERBA MANSA
'onchoshi (V)	'ončoši	*Anemopsis californica*	YERBA MANSA
'ulat' (B, I)	'ulát', 'ulat'	*Juncus acutus*	SPINY RUSH
		Juncus effusus	BOG RUSH
'ush'e'm (V)	'uš'e'm, 'uš'em	water bottle	basket type
'usha'ak (V)	'uša'ak	*Asclepias* spp.	MILKWEED
'utapikets (V)	'utapikets (cf. 'utapi'qitse "= the dam in Mission Canyon. Means they roasted cacomite and it burned")	*Calochortus* sp. (smaller bulb)	MARIPOSA LILY
'utapits (V)	'utapits	*Calochortus* sp. (larger bulb)	MARIPOSA LILY
ch'elhe' tsqono (O)	čeɨhe' tsqono (JPH: lit. 'lengua del venado' = 'deer's tongue')	*Anemopsis californica*	YERBA MANSA, SWAMP ROOT
ch'okoko (O)	č'okoko	*Heteromeles arbutifolia*	TOYON, CHRISTMAS BERRY
cha' (I)	č^{h}a'	*Marah macrocarpus*	WILD CUCUMBER
chatɨshwɨ 'i khus (V)	čatɨšwɨ 'ixus	*Rhamnus californica*	COFFEE BERRY

Popularized Spelling	Linguistic Spelling	Scientific Name	Common Name
chatishwɨ 'ikhshap (V)	č'atɨšwɨ 'ixšap	*Daucus pusillus*	RATTLESNAKE WEED
chaya (B)	č'aya, ča'ya, chaya	leaching tray	basket type
chikwɨ' (B)	čikwɨ'	*Anagallis arvensis*	SCARLET PIMPERNEL
		Erodium spp.	FILAREE
chiqwi 'ikhakha'kh (B)	čiqwí 'ixaxá'x,	*Horkelia cuneata*	HORKELIA
	siqwí 'ixaxá'x	*Potentilla glandulosa*	CINQUEFOIL
chkapsh (V)	čkapš	*Phyllospadix torreyi*	SURF-GRASS, SEAGRASS
chlumes (V)	člumes	*Yucca whipplei*	YUCCA (dry stalk)
chmishɨ (O)	čmišɨ	*Heteromeles arbutifolia*	TOYON, CHRISTMAS BERRY
chnuy (I)	čnuy	*Chlorogalum pomeridianum*	SOAPROOT (fiber)
choch (V)	čoč	*Chenopodium californicum*	SOAP PLANT
chpa' (V)	čp^{h}a', šp^{h}a'	*Lomatium californicum*	CHUCHUPATE
chtɨmɨy (V)	čtɨmɨy	*Ribes amarum*	BITTER GOOSEBERRY
chto (O)	čto	*Prunus ilicifolia*	HOLLY-LEAVED CHERRY
chtu 'ima (V)	čtu 'ima	*Solidago californica*	GOLDENROD
chtu 'iqonon (V)	čtu 'iqonon	*Symphoricarpos mollis*	SNOWBERRY
		Lonicera subspicata var. *denudata*	CHAPARRAL HONEYSUCKLE
helek' (V)	helek'	conical burden basket	basket type
hukhminash (B, I)	huxmínaš, huxminaš	mashed Juniper seed cones	food
k'á'nay (B)	k'á'nay	*Plagiobothrys nothofulvus*	POPCORN FLOWER
kawɨyɨsh (V)	kawɨyɨš	*Scirpus acutus*	BULRUSH, TULE
		Scirpus californicus	CALIFORNIA BULRUSH
kepey (Cr)	kepey	*Dryopteris arguta*	COASTAL WOOD FERN
		Pentagramma triangularis	GOLDBACK FERN
		Polypodium californicum	CALIFORNIA POLYPODY
kepeye (B, V)	kepéye, kepeye	*Dryopteris arguta*	COASTAL WOOD FERN
		Pentagramma triangularis	GOLDBACK FERN
		Polypodium californicum	CALIFORNIA POLYPODY
kh'a'w (B)	x'a'w	*Calochortus (luteus?)*	MARIPOSA LILY
kh'i'm (B, I, V)	x'i'm, x'im, q'i'm	large storage basket	basket type
kh'omho (V)	x'omho, qomho	narrow-mouthed basket	basket type
khakhpanɨsh (P)	kaxpanɨš	*Quercus* spp.	acorns
khaksh (B, I)	xaxš	*Lepidium nitidum*	PEPPERGRASS
khap (V)	xap	*Typha* spp.	CATTAIL
khapshikh (B, I, V)	xapšɨx	*Salvia apiana*	WHITE SAGE

khaw (Cr, V)	xaw, qaw	*Salix lasiolepis*	ARROYO WILLOW
khɨ' (I)	xɨ'	*Opuntia* spp.	PRICKLY-PEAR
khɨ'ɨl (V)	xɨ'ɨl	*Opuntia* spp.	PRICKLY-PEAR
khɨkhɨ' (B)	xɨxɨ'	*Opuntia* spp.	PRICKLY-PEAR
khman (B)	xmán, xmána, man	*Malacothamnus fasciculatus*	CHAPARRAL MALLOW
kholowush (I)	xolowuš	balls of thick acorn mush	food
khsho' (B, V)	xšo'	*Platanus racemosa*	WESTERN SYCAMORE
khutash (B, I, V)	xutaš, xútaš	*Calandrinia* spp.	RED MAIDS
khwapsh (B, I, V)	xwapš, xwapč	*Urtica dioica* subsp. *holosericea*	GIANT CREEK NETTLE
khwelekhwel (V)	xwelexwel	*Populus* spp.	COTTONWOOD
kich (V)	kič	*Pteridium aquilinum*	BRACKEN FERN
kɨwɨkɨw (V)	kɨwɨkɨw	*Equisetum* spp.	HORSETAIL
		Ephedra spp.	DESERT / MOUNTAIN TEA
kot' (I)	kot'	*Chlorogalum pomeridianum*	SOAP PLANT, SOAPROOT
ktɨp (B)	ktɨp	*Juglans californica*	BLACK WALNUT
ku'w (B, I)	ku'w	*Quercus agrifolia*	COAST LIVE OAK
kupushash (B)	kupušaš, kupuša'aš	drinking cup	basket type
kuw (V)	kuw	*Quercus agrifolia*	COAST LIVE OAK
kuwu (Cr)	kuwu	*Quercus agrifolia*	COAST LIVE OAK
kuyiwash (V)	kuyiwaš, kuyiwas	basketry dish	basket type
kwɨ'ɨn (V)	kwɨ'ɨn	*Erodium* spp.	FILAREE
l'ɨpɨ (O)	l'ɨpɨ	*Salvia columbariae*	CHIA
lilak (V)	lilak, lilak^{h}	*Croton californicus*	CALIFORNIA CROTON
lit'on (I)	lit'on, malit'on	*Distichlis spicata*	SALT GRASS
makhsik (V)	maxsik	*Clematis* spp.	CLEMATIS
makhsik' (B, P)	maxsik'	*Clematis* spp.	CLEMATIS
mal (B, I)	mal	*Malva nicaeensis, M. parviflora*	MALLOW, CHEESEWEED
malwash (V)	malwaš	*Malva nicaeensis, M.parviflora*	MALLOW, CHEESEWEED
man (B)	man, xmán, xmána	*Malacothamnus fasciculatus*	CHAPARRAL MALLOW
manakhshmu (V)	manaxšmu	*Helenium puberulum*	SNEEZEWEED
mashqupshlét 'akhɨkhaw (I)	mašqupšlét' 'axuxa'w	*Castilleja* spp.	INDIAN PAINTBRUSH
masteleq 'a pistuk (I)	masteleq 'apistuk	*Achillea millefolium*	YARROW
matakh (B)	mátax	*Calochortus (venustus?)*	MARIPOSA LILY
mekhme'y (B, I, V)	mexme'y	*Juncus textilis*	INDIAN RUSH
meq (B,I)	meq	*Yucca whipplei*	YUCCA (dry stalk)
meqme'i (Cr)	meq^{h}me'i	*Juncus textilis*	INDIAN RUSH
mim (B, I, V)	mim, mi'm	*Paeonia californica*	PEONY
mish'kata (P)	miš'kata	*Quercus douglasii*	BLUE OAK

Popularized Spelling	Linguistic Spelling	Scientific Name	Common Name
mɨs (B, I, V)	mɨs	*Quercus berberidifolia, Q. dumosa*	SCRUB OAK
mɨsɨ (Cr)	mɨsɨ	*Quercus* spp.	acorns
mo’ (V)	mo’	*Atriplex* spp.	SALTBUSH
mo’kh (B, I)	mo’x	*Cucurbita foetidissima*	WILD GOURD
mo’moy (B, Cr)	mó’moy, mo’moy	*Datura wrightii*	JIMSONWEED, TOLOACHE
mo’okh (V)	mo’ox	*Cucurbita foetidissima*	WILD GOURD
mo’yoq’ (B)	mo’yoq’	*Zigadenus* spp.	ZYGADENE, DEATH CAMAS
mokoksh (I)	mokokš	*Xanthium strumarium*	COCKLEBUR
molɨsh (V)	molɨsh	*Artemisia douglasiana*	MUGWORT
molo’wot’ (B)	moló’wot’	*Marah macrocarpus*	WILD CUCUMBER
molush (B, I)	moluš	*Artemisia douglasiana*	MUGWORT
momoy (V, I)	momoy	*Datura wrightii*	JIMSONWEED, TOLOACHE
mow (B, I, V)	mow	*Alnus rhombifolia*	WHITE ALDER
moyoq (V)	moyoq	*Zigadenus* spp.	ZYGADENE, DEATH CAMAS
moyoq’ (I)	moyoq’	*Zigadenus* spp.	ZYGADENE, DEATH CAMAS
mulus (B, I, V)	múlus, mulus	*Juniperus californica*	CALIFORNIA JUNIPER
na’ (B, I, P)	na’	*Adenostoma fasciculatum*	CHAMISE
nakhaykha’y (B)	naxayxá’y	*Calochortus (catalinae?)*	MARIPOSA LILY
nu’nit’ (B)	nu’nit’	*Vitis californica, V. girdiana*	WILD GRAPE
nunit (V)	nunit	*Vitis californica, V. girdiana*	WILD GRAPE
nunit’ (I)	nunit’	*Vitis californica, V. girdiana*	WILD GRAPE
pa’ (B, I)	p^{h}a’	*Lomatium californicum*	CHUCHUPATE
pakh (I,V)	pax	*Salvia carduacea*	THISTLE SAGE
pakh (V)	pax	*Salvia spathacea*	HUMMINGBIRD SAGE
pash (V)	paš	*Chlorogalum pomeridianum*	SOAP PLANT, SOAPROOT
peye (I)	peye	*Dryopteris arguta*	COASTAL WOOD FERN
		Pentagramma triangularis	GOLDBACK FERN
		Polypodium californicum	CALIFORNIA POLYPODY
pɨ (V)	pɨ	walnut-shell dice game	game type
pɨch (B)	pɨč	*Cercocarpus betuloides*	MOUNTAIN-MAHOGANY
pɨtsmu (V)	pɨtsmu	tarred canoe bailer	basket type
pokh (B, I)	pox	*Yucca whipplei*	YUCCA (plant, leaves)
posh (B, I, V)	poš	*Pinus monophylla*	PINYON
psha’an (V)	pša’an	*Umbellularia californica*	CALIFORNIA BAY, LAUREL
psha’n (B, I)	pša’n, pča’n	*Umbellularia californica*	CALIFORNIA BAY, LAUREL
pulash (I)	pulaš	*Calochortus* sp.	MARIPOSA LILY

puq' (B, I)	puq'	*Rhamnus californica*	COFFEE BERRY
qayas (B, Cr, I, V)	qayas, qáyas	*Sambucus mexicana*	BLUE ELDERBERRY
qayɨsh (B, I)	qayɨš	*Cirsium* spp.	THISTLE
		Sonchus oleraceus	COMMON SOW-THISTLE
		Hypochaeris radicata	HAIRY CAT'S EAR
qi'w (B)	qi'w	*Chlorogalum pomeridianum*	SOAP PLANT, SOAPROOT
qi'w (B, V)	qi'w	seedbeater	basket type
qimsh (B)	qimš	*Salvia spathacea*	HUMMINGBIRD SAGE
qɨch (B)	qɨč	*Pteridium aquilinum*	BRACKEN FERN
qlaha' (B)	qlahă'	*Lupinus* spp.	LUPINE
qlahaw' (V)	qlahaw'	*Lupinus* spp.	LUPINE
qloqol (Cr)	qloqol	*Artemisia douglasiana*	MUGWORT
qolpo'op (Cr, V)	qolpo'op	*Solanum douglasii*	DOUGLAS NIGHTSHADE
qsho' (Cr)	qšo'	*Platanus racemosa*	WESTERN SYCAMORE
qupe (B, I, V)	qúpe, qupe	*Eschscholzia californica*	CALIFORNIA POPPY
qwap'sh (Cr)	qwap'š	*Urtica dioica* subsp. *holosericea*	GIANT CREEK NETTLE
qwe (V)	qwe	*Heteromeles arbutifolia*	TOYON, CHRISTMAS BERRY
qwe' (B, Cr, I)	qwe', qwé'	*Heteromeles arbutifolia*	TOYON, CHRISTMAS BERRY
qweleqwe'l (B)	qweleqwe'l, qweɨqwe'l	*Populus* spp.	COTTONWOOD
qweleqwel (I)	qweleqwel	*Populus* spp.	COTTONWOOD
s'akhiyɨp 'ikhshap (B)	s'axi'yɨp 'ixšap	*Daucus pusillus*	RATTLESNAKE WEED
s'akht'utun 'iyukhnuts (B)	s'axt'utun 'iyuxnuts, s'axtutul yuxnuts (descriptive: 'hummingbird sucks it')	*Epilobium canum*	CALIFORNIA FUCHSIA
s'akhtutu 'iyukhnuts (B)	s'axtutu 'iyuxnuts (descriptive: 'hummingbird sucks it')	*Silene laciniata*	INDIAN PINK
s'epsu' 'aquqa'w (I)	s'epsu' 'aquqa'w, s'aqtaku' 'aquqa'w (lit. 'it's-hat the-coyote' = 'coyote's hat')	*Calystegia macrostegia*	MORNING-GLORY
s'epsu 'i'ashk'á' (B)	s'epsu' hul'ašk'a', s'epsu' 'i'ašk'á' (lit. 'it's-hat the-coyote' = 'coyote's hat')	*Calystegia macrostegia*	MORNING-GLORY
s'u'wlima' (B, I)	s'u'wlima' (s-'uw hi ma': 'its-food the-jackrabbit' = 'jackrabbit's food')	*Erodium* spp.	FILAREE

Popularized Spelling	Linguistic Spelling	Scientific Name	Common Name
s'uwmo' 'oyoso (I)	s'uwmo' 'oyoso (lit. 'its-food the-bee' = 'bee's food')	*Verbena lasiostachys*	WESTERN VERVAIN
saha (V)	saha	*Distichlis spicata*	SALT GRASS
sekh (B, I, V)	sex	*Ceanothus megacarpus*	BIGPOD CEANOTHUS
sh'ichkɨ 'i'waqaq (B)	š'ičkɨ 'i'waqaq, ts'itsk'í 'iwaqáq	*Sisyrinchium bellum*	BLUE-EYED GRASS
sha'puk' (B)	ša'púk'	*Trifolium* spp.	CLOVER
sha'w (I)	ša'w	*Rumex hymenosepalus*	CANAIGRE
shakh (I, V, Cr)	šax	*Leymus condensatus*	GIANT RYE
shapuk (V)	šapuk	*Trifolium* spp.	CLOVER
shapuk' (I)	šapuk'	*Trifolium* spp.	CLOVER
shaw (B)	šaw	*Rumex hymenosepalus*	CANAIGRE
shi'q'o (V)	ši'q'o	*Dichelostemma capitatum*	BLUE DICKS, BRODIAEA
shikhwapsh 'i'ashk'a' (B, V)	šixwapš 'i'ašk'á' (lit. 'its-nettle the-coyote' = 'coyote's nettle')	*Verbena lasiostachys*	WESTERN VERVAIN
shilik (I)	šilik	*Claytonia perfoliata*	MINER'S LETTUCE
shilik' (B)	šilik'	*Claytonia perfoliata*	MINER'S LETTUCE
shipitish (V)	šipitiš	acorn mush	food
shiq'o'n (B, I)	šix'ó'n, šix'o'n	*Dichelostemma capitatum*	BLUE DICKS, BRODIAEA
shiyamsh 'ikhshap (I)	šiyamš 'ixšap	*Daucus pusillus*	RATTLESNAKE WEED
shɨpɨtɨsh (B, I)	šɨpɨtɨš	acorn mush	food
shkash (B)	škaš	*Phyllospadix torreyi*	SURF-GRASS, SEAGRASS
shlamulasha'w (B)	šlamulaša'w	*Phoradendron* spp.	MISTLETOE
shnokhsh'en (I)	šnoxš'en	*Yucca whipplei*	YUCCA (stalk tip, esp. cooked)
sho'moy (B)	šo'moy	*Xanthium strumarium*	COCKLEBUR
shomoy (V)	šomoy, čomoy	*Xanthium strumarium*	COCKLEBUR
shonush (I)	šonuš, xšonuš	*Platanus racemosa*	WESTERN SYCAMORE
show (B, I, V)	šow	*Nicotiana* spp.	TOBACCO
shtakuk (Cr, V)	štakuk, čtakuk	*Yucca whipplei*	YUCCA (plant, stalk)
shtamhɨl (V)	štamhɨl, čtamhɨl, štamxɨl	*Carpobrotus chilensis*	SEA FIG
shtayɨt (B, I, P)	štayɨt, stayɨt	*Salix lasiolepis*	ARROYO WILLOW
shtemele (P)	štemele	*Leymus condensatus*	GIANT RYE
shtemelel (B)	štemélel, temelel	*Leymus condensatus*	GIANT RYE

shtop'o hul'alaqshan (B)	štop'o huɬ'aqšan (lit. 'its-navel that-dead-one' = 'dead man's navel')		mushroom, fungi
shtopo' 'alaqshan (I)	štopo' 'alaqšan (lit. 'its-navel that-dead-one' = 'dead man's navel')		mushroom, fungi
shtopo'y 'i shakshanuch (V)	štopo'y 'išakšanuč (lit. 'its-navel that-dead-one' = 'dead man's navel')		mushroom, fungi
shtoyho'os (V)	štoyho'òs	*Carpobrotus chilensis*	SEA FIG
shtoyho'os (V)	štoyho'os, tstoyho'os	*Rhus integrifolia*	LEMONADEBERRY
		Rhus ovata	SUGAR BUSH
shtu'ama' (I)	štu'ima, stu'ama'	*Solidago californica*	GOLDENROD
shu' (B, I)	šu'	*Baccharis salicifolia*	MULE FAT, WATER WALLY
shu'nay (B, I)	šu'nay	*Rhus trilobata*	THREE-LEAVED SUMAC
shukepesh 'ishaq (V)	šukepeš 'išax	*Arundo donax*	GIANT REED
shuna'y (V)	šuna'y	*Rhus trilobata*	THREE-LEAVED SUMAC
shushtewesh (I)	šušteweš, mašušteweš (descriptive: 'sharp, a spine that digs into your skin')	*Avena* spp.	WILD OAT
sihi (B)	sihi	*Amaranthus* spp.	AMARANTH, PIGWEED
sihi' (V)	sihi'	*Amaranthus* spp.	AMARANTH, PIGWEED
smakhna'atl (V)	smaxna'atɬ	*Croton californicus*	CALIFORNIA CROTON
snokhsh'en (V)	snoxš'en	*Yucca whipplei*	YUCCA (stalk tip, esp. cooked)
spe'ey he'so'o (V)	spe'ey he'so'o (descriptive: 'flower of the water')	*Rorippa nasturtium-aquaticum*	WATERCRESS
sq'o'yon (I)	sq'o'yon	*Arctostaphylos* spp.	MANZANITA
sq'oyon (B)	sq'óyon	*Arctostaphylos* spp.	MANZANITA
sqa'yi'nu (B)	sqa'yi'nu	*Ribes* spp.	CURRANT
sqayi'nu (I)	sqayi'nu	*Ribes* spp.	CURRANT
stakuk (B, I)	stakuk, stakúk	*Yucca whipplei*	YUCCA (stalk)
stapan (B, Cr)	stapan, štapan, čtapan	*Scirpus acutus*	BULRUSH, TULE
stapan (B, Cr, I)	stapan, štapan, čtapan	*Scirpus californicus*	CALIFORNIA BULRUSH
ste'leq' 'ipistuk' (B)	ste'leq' 'ipistuk' (lit. 'its-tail the-squirrel' = 'squirrel's tail')	*Castilleja* spp.	OWL'S CLOVER

Popularized Spelling	Linguistic Spelling	Scientific Name	Common Name
stelek 'ipistúk (I)	stelek 'ipistúk (lit. 'its-tail the-squirrel' = 'squirrel's tail')	*Castilleja* spp.	OWL'S CLOVER
steleq' 'a'emet (I)	steleq' 'a'emet (lit. 'its-tail the-ground-squirrel' = 'ground-squirrel's tail')	*Achillea millefolium*	YARROW
stɨmɨy (B, I)	stɨmɨy, tstɨmɨy	*Ribes amarum*	BITTER GOOSEBERRY
stɨmɨy 'iwɨ (B)	stɨmɨy 'iwɨ (lit. 'its-gooseberry the-deer' = 'deer's gooseberry')	*Ribes speciosum*	GOOSEBERRY
stɨq' 'iwaq'aq' (B)	stɨq 'iwaq'aq (lit. 'its-eye the-frog' = 'frog's eye')	*Dodecatheon clevelandii*	SHOOTING STAR
stɨq shi'sha'w (V)	stɨq ši'ša'w	*Grindelia camporum*	GUM PLANT
sto'yots' (B, I)	stó'yots', sto'yots'	*Carpobrotus chilensis*	SEA FIG
stu 'imá (B)	stú' 'imá'	*Solidago californica*	GOLDENROD
stu'yi' (O)	stu'yi, tstuyi'	*Nicotiana* spp.	TOBACCO
stumuku'n (I)	stumuku'n	*Phoradendron* spp.	MISTLETOE
su'núk' (B)	su'núk'	*Chenopodium californicum*	SOAP PLANT
swa' (B, I)	swa'	*Scirpus americanus*, *S. pungens*	THREE-SQUARE BULRUSH
swey (B, I, P, V)	swéy, swey (lit. 'it is open, it gaps')	*Hemizonia fasciculata*	TARWEED
swo's 'i'ashk'á' (B)	swo's 'i'ašk'á', swo's hulašká', siwamɨs 'i'ašká'	*Castilleja* spp.	INDIAN PAINTBRUSH
swow (I, P)	swow, swo'w	*Scirpus acutus*	BULRUSH, TULE
syɨt (V)	syɨt	*Juncus textilis* (stem's red base)	INDIAN RUSH
ta (V)	ta	*Quercus lobata*	VALLEY OAK
ta' (B, I, Cr, O)	ta', ta'a	*Quercus lobata*	VALLEY OAK
tak (B, I)	tak	*Pinus* spp.	PINE (several kinds)
taqsh (B, I)	taqš, takč	*Typha* spp.	CATTAIL
tash (V)	taš	*Juncus balticus*	WIRE RUSH
		Juncus textilis (smaller stems)	INDIAN RUSH
teqep'sh (I)	teqep'sh	seed beater	basket type

tekhe'we (B)	texe'we	*Cryptantha intermedia*	CRYPTANTHA
		Amsinckia menziesii	FIDDLENECK
tekhewe' (I)	texewe'	*Amsinckia menziesii*	FIDDLENECK
		Cryptantha intermedia	CRYPTANTHA
teksu (O)	teksu, texsu	*Platanus racemosa*	WESTERN SYCAMORE
temala (O)	temala	*Malva nicaeensis, M. parviflora*	MALLOW, CHEESEWEED
tenech (B)	teneč	*Keckiella cordifolia*	CLIMBING PENSTEMON
		Lonicera subspicata var. *subspicata*	S. B.HONEYSUCKLE
tilho (O)	tiłho	*Artemisia californica*	COASTAL SAGEBRUSH
tipaq' (I)	tipaq', tipax	large tray	basket type
tɨhɨ (V)	tɨhɨ	*Rubus ursinus*	CALIFORNIA BLACKBERRY
tɨpk (V)	tɨpk	*Juglans californica*	BLACK WALNUT
tɨq'ɨtɨq' (B)	tɨq'ɨtɨq'	*Rubus ursinus*	CALIFORNIA BLACKBERRY
tɨqɨtɨq (I)	tɨqɨtɨq	*Rubus ursinus*	CALIFORNIA BLACKBERRY
tluyash (Cr)	tłuyaš	acorn mush	food
tok (B, I)	tok	*Apocynum cannabinum*	INDIAN-HEMP
tok (Cr)	tok, tok^{h}u, patok^{h}u	*Apocynum cannabinum*	INDIAN-HEMP
tok (V)	tok, tok^{h}u	*Apocynum cannabinum*	INDIAN-HEMP
tomol (B, I, V)	tomol	boat, plank canoe	watercraft
tomol (B, I)	tomol	*Pinus jeffreyi, P. ponderosa*	PINE (certain kinds)
topo (V)	topo	*Phragmites australis*	CARRIZO GRASS
t'pi'nɨ (O)	t'pi'nɨ	*Juniperus californica*	CALIFORNIA JUNIPER
tpinusmu' (O)	tpinusmu'	*Artemisia douglasiana*	MUGWORT
tpɨtɨ (O)	tpɨtɨ	*Quercus* spp.	acorns
tqɨ'ɨ (O)	tqɨ'ɨ	*Opuntia* spp.	PRICKLY-PEAR
tqmapsɨ (O)	tqmapsɨ	*Urtica dioica* subsp. *holocericea*	GIANT CREEK NETTLE
tqmimu' (O)	tqmimu'	*Leymus condensatus*	GIANT RYE
tqupa' (O)	tqupa'	*Chlorogalum pomeridianum*	SOAPROOT, SOAP PLANT
ts'aqsmi (V)	ts'aqsmi	*Cirsium* spp.	THISTLE
		Hypochaeris radicata	HAIRY CAT'S EAR
		Sonchus oleraceus	COMMON SOW-THISTLE
ts'aqwɨpwɨp (I)	ts'aqwɨpwɨp	*Salix exigua?*	SANDBAR WILLOW
ts'isuyi' (O)	ts'isuyi'	*Yucca whipplei*	CHAPARRAL YUCCA
ts'ukhat' (I)	ts'uxat'	*Rumex crispus*	CURLY DOCK
tsa' (O)	tsa'	*Salix lasiolepis*	ARROYO WILLOW
tsaya (I, V)	tsaya	leaching tray	basket type
tsiqun (I)	tsiqun	*Ribes speciosum*	GOOSEBERRY

Popularized Spelling	Linguistic Spelling	Scientific Name	Common Name
tsɨkɨnɨn (V)	tsɨk^{h}ɨnɨn	*Pinus* sp.	WHITE PINE (general)
tspasɨsɨ (O)	tspasɨsɨ	*Allium* spp.	WILD ONION
tsqoqo'n (V)	tsqoqo'n	*Arctostaphylos* spp.	MANZANITA
tsukh	tsux	topknot headdress	dance regalia
tsukhat' (B)	tsuxat', stuxat'	*Rumex crispus*	CURLY DOCK
tsuwu' (O)	tsuwu'	*Quercus agrifolia*	COAST LIVE OAK
tswana'atl 'ishup (V)	tswana'atɬ heešup	*Eriogonum fasciculatum*	CALIFORNIA BUCKWHEAT
tu' (B, I)	tu'	*Symphoricarpos mollis*	SNOWBERRY
		Lonicera subspicata var. *denudata*	CHAPARRAL HONEYSUCKLE
tup (V)	tup	*Scirpus americanus, S. pungens*	THREE-SQUARE BULRUSH
tushqun (I)	tušqun	*Quercus douglasii*	BLUE OAK
twalili' (O)	twalilí'	*Baccharis salicifolia*	MULE FAT, WATER WALLY
wachka'y (I)	wačka'y, wačkay	gathering basket	basket type
wak (B, Cr, I, V)	wak	*Salix laevigata*	RED WILLOW
wala (O, P)	wala, twala	*Toxicodendron diversilobum*	POISON-OAK
wala'laq' (B)	wala'laq	*Lupinus* spp.	LUPINE
walqaqsh (B, I)	walqaqš	*Rhus integrifolia*	LEMONADEBERRY
		Rhus ovata	SUGAR BUSH
walqaqsh (B, I, V)	walqaqš	*Malosma laurina*	LAUREL SUMAC
wam (Cr)	wam	*Prunus ilicifolia*	HOLLY-LEAVED CHERRY
washiko (B, I)	wašiko	*Ceanothus oliganthus*	HAIRY CEANOTHUS
		Ceanothus spinosus	GREENBARK CEANOTHUS
washiko (V)	wašiko, watik'o	*Ceanothus oliganthus*	HAIRY CEANOTHUS
		Ceanothus spinosus	GREENBARK CEANOTHUS
washtiq'oliq'ol (B, I)	waštiq'oliq'ól, wačtiq'oliq'ol	*Rosa californica*	CALIFORNIA WILD ROSE
watik (V)	watik	boiling basket	basket type
watiq'oniq'on (V)	watiq'oniq'on	*Rosa californica*	CALIFORNIA WILD ROSE
we'lel (B)	we'lel	*Chenopodium berlandieri*	PITSEED GOOSEFOOT
we'wey (B)	we'wey	*Artemisia californica*	COASTAL SAGEBRUSH
welel (I,V)	welel	*Chenopodium berlandieri*	PITSEED GOOSEFOOT
welu (B)	welu (loan word: from Sp. berro, 'watercress')	*Rorippa nasturtium-aquaticum*	WATERCRESS
wewe'y (V)	wewe'y	*Artemisia californica*	COASTAL SAGEBRUSH
wewey (Cr, I)	wewey	*Artemisia californica*	COASTAL SAGEBRUSH

wi'ma (B)	wi'ma	*Sequoia sempervirens*	COAST REDWOOD
wili'lik' (B)	wili'lik'	*Baccharis plummerae*	PLUMMER'S BACCHARIS
		Conyza canadensis	HORSEWEED
wililik (V)	wililik	*Baccharis plummerae*	PLUMMER'S BACCHARIS
		Conyza canadensis	HORSEWEED
wililik' (I)	wililik'	*Baccharis plummerae*	PLUMMER'S BACCHARIS
		Conyza canadensis	HORSEWEED
wiliq'ap' (B)	wiliq'ap'	*Cornus* spp.	DOGWOOD
wiliqap (V)	wiliqap	*Cornus* spp.	DOGWOOD
wima (V)	wima, wimal	*Sequoia sempervirens*	COAST REDWOOD
wima' (I)	wima'	*Sequoia sempervirens*	COAST REDWOOD
wiqap' (I)	wiqap'	*Cornus* spp.	DOGWOOD
wisay (I)	wisay	fish trap	fishing equipment
wishap (I, V)	wiš^{h}ap	*Eriodictyon crassifolium*	YERBA SANTA
wishap' (B)	wiš^{h}áp'	*Eriodictyon crassifolium*	YERBA SANTA
wita'y (V)	wita'y	*Baccharis salicifolia*	MULE FAT, WATER WALLY
wɨ'lɨ (V)	wɨ'lɨ	*Lyonothamnus floribundus*	ISLAND IRONWOOD
wɨlɨ (Cr)	wɨlɨ	*Lyonothamnus floribundus*	ISLAND IRONWOOD
wɨltɨ'y (V)	wiltɨ'y	*Fraxinus dipetala*	FOOTHILL ASH
wɨntɨ'y (B, I)	wɨntɨ'y	*Fraxinus dipetala*	FOOTHILL ASH
wɨp (B, I, V)	wɨp	*Yucca whipplei*	YUCCA (crown, esp. cooked)
wo'ni (V)	wo'ni	burden basket	basket type
woshk'o'loy (B)	wošk'ó'loy,	*Ephedra* spp.	DESERT / MOUNTAIN TEA
	woskq'ó'loy,	*Equisetum* spp.	HORSETAIL
	wošk'ó'lol		
ya'i (B)	ya'y	*Lotus scoparius* (?)	DEERWEED
yasis (B, Cr, I, V)	yasis, yásis	*Toxicodendron diversilobum*	POISON-OAK
yai (V)	yay	*Lotus scoparius* (?)	DEERWEED
yep (B)	yep	tarred parching tray	basket type
yepunash (B, V)	yepúnasa, yepúnaš	*Achillea millefolium*	YARROW
yɨw (V)	yɨw	winnowing tray	basket type

PLANT PARTS

Acorn	sk'uk'uy (I)
Bark, its bark	s^{h}ol (B)
Bark (rough), husk	s^{h}ol (I)
Bark (smooth)	pax (I)
Bark, nutshell	popo (O)
Brush, bush	tɨp (B)
Brush, chaparral	tɨp (I)
Burl	spukuy (I)
Burl; pine cone	sko'w (I)
Catkin of live-oak	swawáyan (B)
Catkin of willow	sqapuni (I)
Fallen leaves	štoxosxos (I)
Flower	spey, spe'y, spe'ey (B)
Flower	spey (I)
Flower	tšpe'^{e} (O)
Grass	-pšelpše (O)
Greens	'anɨtɨš (B)
Herb, grass, any plant	tsweq (I)
Herb, medicine	tšatɨswɨ (V)
Herb, medicine, vegetable	'axulapšan (I)
Hollow of a tree	qalawáwaq (B)
Leaf, its leaf	sqap (B)
Leaf	sqap (I)
Leaf	tu'u (O)
Leaf blight	štixup (I)
Log, a big log	syulok'in (I)
Medicine	'alxulapšan (B)
Pit, hard seed	toqos (B)
Pitch	spɨł (B)
Pith	'ayapis (I)
Pollen	šnuy (I)
Root	s'axpilil(B)
Root, its root	s'axpilɨ'l (I)
Root	qspiłhi (O)
Sap	s'o'o (B)
Sap, juice	šapš (V)
Sap, plant juice	š'apš (I)
Seed	'amɨn (I)
Seed pod	'amin (B)
Seed pod	sxelexeḷ (I)
Seed, fruit	ts'imɨ'^{ɨ} (O)
Seed, pit of fruit	tokos (I)
Shoot	steneyep'un (B)
Shoot	šlumes (I)
Skin (of fruit), smooth bark	spax (B)
Skin, peel of a fruit	šlɨw (I)
Stamens	sqɨnqɨnɨl (B)
To sprout	tilɨk (I)
Tree	pono'o, tpono (O)
Tree, stick	pon (I)
Tree, wood	pon (B)
Trunk	stɨpɨq (B)

INDEX

B

C

G

H

I

J

N

O

T

ABOUT THE AUTHOR

Photograph by Rose Briccetti

JAN TIMBROOK

Jan Timbrook, Ph.D., is Curator of Ethnography at the Santa Barbara Museum of Natural History, where she has been on the Anthropology staff since 1974. She is recognized as one of the top experts on Chumash life and culture, and this book is the result of some thirty years of research.

Timbrook's wide-ranging published writings also address environmental management, herbal medicine, the plank canoe, basketry, birds in Chumash culture, and more. She is particularly interested in weaving of various kinds, especially California Indian basketry. She contributed a chapter on caring for ethnographic materials to an authoritative manual for museum professionals on curation of biocultural collections.

In addition to her scientific contributions, Timbrook shares her knowledge with the wider public and with the peoples who made the objects she studies. She gives many presentations to community groups each year and prepares various exhibitions of North American Indian cultural materials. She and her botanist husband live in Santa Barbara, not far from the museum.